GROUND TREATMENT BY DEEP COMPACTION

THE INSTITUTION OF CIVIL ENGINEERS
LONDON 1976

Published and distributed by Thomas Telford Ltd
for the Institution of Civil Engineers at 26–34 Old
Street, London EC1V 9AD.

The papers in this volume were first published as a
Symposium in Print in Géotechnique March 1975.

ISBN 0 7277 0024 3

CONTENTS

Preface

The purpose of this symposium is to bring together a collection of papers on a subject of practical importance in foundation engineering. Much has been written on the subject of deep compaction by vibratory methods. Nearly all the papers have been produced by contractors specializing in these techniques and, not unnaturally, they have concentrated on the successes obtained by their methods. The failures, or at least the lack of apparent successful applications, have remained unrecorded. The result of this has been the growth of a certain mystique surrounding the techniques, and claims have been made on their ability to 'strengthen' ground which cannot always be substantiated when subjected to a critical review.

Papers in the symposium include some by authors from the specialist firms exploiting these techniques, some from authors with experience of their practical effect from the point of view of the user, and some from research workers in the field. It is hoped that it represents a broadly-based survey of research, practical application and user experience.

The compaction methods dealt with fall into three main divisions: vibro-compaction of non-cohesive soils, stone columns in cohesive soils and dynamic consolidation.

The paper by Thorburn presents a general review of the subject and gives useful design rules for allowable bearing pressures on cohesionless soils and granular fills compacted by deep vibration equipment, and allowable loads on stone columns formed in weak cohesive soils. The paper by West includes a comprehensive table setting out the types of structure and ground conditions which are favourable or otherwise for deep compaction techniques. Both West and Rathgeb and Kutzner give examples of the successful use of deep vibro-compaction techniques for the founding of important structures on loose or medium dense granular soils. Abelev gives a detailed account of the techniques used for the deep compaction of loess soils; these soils present serious foundation problems in the USSR and other parts of Eastern Europe. The compaction techniques are simple and would appear to be very effective.

While the effectiveness of vibro-compaction in free-draining non-cohesive soils appears to be beyond question the same cannot be said of stone columns in soft cohesive soils. The authors appear to agree that the success of the technique relies on the stone columns deriving their stiffness and strength from the confining action of the surrounding soil. West states that the surrounding clay is unaffected by the action of the vibrator and Thorburn (Fig. 2) suggests that the strength of the soft clay may even be reduced.

The evidence is that stone columns when loaded rapidly perform in accordance with the theory established by the model experiments of Hughes and Withers (*Ground Engineering*, May 1974 7, No. 3, 42–44, 47–49). Hughes, Withers and Greenwood show that when the actual column dimensions are used and the load is applied rapidly over a half-hour period, the predicted performance of a single stone column is in very good agreement with the actual performance. West reports satisfactory performance of stone columns in a soft silty clay as do Engelhardt and Golding. However, the question must be raised as to whether or not the supporting action of the soft clay can always be relied upon under the slow application of load. McKenna, Eyre and Wolstenholme have presented the settlement records of a 7·9 m high embankment. One end of the embankment was supported on stone columns and the rest was built over untreated ground. The records show that the columns had no effect on either the amount or rate of settlement. McKenna *et al.* suggest that this behaviour resulted from the clay squeezing into the voids of the column thereby offering little or no lateral restraint and also preventing the columns acting as drains.   The explanation is certainly plausible and since this is one of the few reported cases in which a full-scale comparison has been made between treated and untreated ground the possibility of such behaviour must be taken seriously.   It is evident that a considerable amount of work still needs to be done including further full-scale comparative tests on the lines described by McKenna *et al.* before the behaviour of stone columns in soft cohesive ground is adequately understood.

A major benefit of stone columns may be to strengthen soft ground against horizontal loading as applied by either seismic forces or under embankments by fill.   The papers by Rathgeb and Kutzner and Engelhardt and Golding describe cases where stone columns were used for this purpose.   Before employing stone columns for this purpose it may be worth investigating the relative cost of employing a gravel filled trench beneath the toe of the embankment slopes.

The technique of dynamic consolidation offers many attractions. The application of a compaction technique to non-saturated soils and loose granular soils presents few conceptual difficulties and Scott and Pearce present a quantitative theoretical treatment which may be applicable to these soils.   However, there can be no denying that the effectiveness of the technique for soft cohesive soils is difficult to account for by means of simple, classical soil mechanics theory.   Ménard and Broise offer a qualitative explanation of the mechanics of the process.   They consider that the application of intense impact energy to the ground surface induces liquefaction and this, coupled with the formation of cracks, leads to a large temporary increase in mass permeability giving rise to rapid settlement, dissipation in pore-water pressure and gain in strength.   When the cracks close the rate of pore-pressure dissipation reduces markedly as indicated in Fig. 4 of the paper.   The observations from two soft clay sites suggest that the method can be effective but direct comparisons of full-scale behaviour between treated and untreated ground are urgently required.   One factor that may have to be considered is whether the improvement resulting from compaction following the destruction of the clay fabric outweighs the loss of stiffness and strength resulting from such disruptive action.   Thus, care is needed when considering applying the method to sensitive clays.   Also, the length of time required to dissipate the excess pore-pressures after crack closure may be excessive.   As with stone columns the application of dynamic consolidation to soft cohesive soils merits further investigation and field studies.

The papers represent an effort by the Advisory Panel of Géotechnique to provide an up-to-date view of a geotechnical process which is of substantial interest to geotechnical engineers.

J. B. BURLAND
J. M. McKENNA
M. J. TOMLINSON

Theoretical and practical aspects of dynamic consolidation

L. MÉNARD* and Y. BROISE*

This Paper discusses the possibility of improving the mechanical characteristics of fine saturated soils to a great depth by a technique known as dynamic consolidation. The energy required is supplied by repeated high intensity impacts. A review is made of the fundamental research which was carried out to gain an understanding of the mechanisms involved; it treats four main points: (a) the compressibility of saturated soils due to the presence of micro-bubbles; (b) the gradual liquefaction under repeated impacts; (c) the changes of permeability of a soil mass due to the creation of fissures; (d) the thixotropic recovery. The role played by adsorbed water in the last two points is reviewed. The various stages of dynamic consolidation are then summarized in graphical form; the resulting vibrations are also discussed. The control and engineering design parameters are then discussed and the 'dynamic oedometer', a tool specially developed for that purpose, is described. The technique is then illustrated by field results amongst which was one in Boston (UK) where 'buttery' clay was treated.

La consolidation dynamique permet l'amélioration des propriétés mécaniques des sols fins saturés jusqu'à des profondeurs importantes. L'énergie requise est transmise au terrain par des impacts répétés de forte intensité. La recherche fondamentale qui a dû être entreprise, afin de comprendre les divers phénomènes associés au procédé, est examinée; elle traite 'essentiellement de quatre points: (a) la compressibilité de sols saturés dûe à la présence de micro-bulles; (b) la liquéfaction progressive sous des impacts répétés; (c) les variations de la perméabilité d'un sol qui résultent de la fissuration dans la masse; (d) le regain thixotropique. Le rôle important que joue l'eau adsorbée sur ces deux derniers points est expliqué. Les différents effets de la consolidation dynamique sont présentés sous forme de graphiques et les effets des vibrations induites sont expliqués brièvement. Les paramètres nécessaires au contrôle et à la réalisation technique sont examinés et 'l'œdomètre dynamique', instrument développé pour les maîtriser, est décrit. Enfin, plusieurs résultats de chantier sont présentés y compris un à Boston (GB) où de l'argile très plastique a été traitée.

In 1970, Techniques Louis Ménard introduced a technique initially known as heavy tamping. Its field of application then covered principally ballast fills or natural sandy gravel soils; it was soon found possible, however, to extend this field to saturated clay or alluvial soils. From then onwards the technique has taken the name 'dynamic consolidation'.

With dynamic consolidation, the improvement in the mechanical characteristics of a soil which is compressible for a considerable depth (10–30 m) is obtained by the repeated application of very high intensity impacts to the surface. The procedure consists of dropping pounders weighing tens of tonnes from great heights (15–40 m) while following a well defined programme as regards time and space appropriate to the site. The technique and

---

*Ménard Techniques Limited, Aylesbury.

some of the instrumentation developed for its control are covered by patents and trade names in most industrialized countries.

The dynamic consolidation technique rapidly gained acceptance for the treatment of coarse or non-saturated soils. However, as soon as it was applied to clayey soils, marked scepticism was encountered in spite of the consistently favourable results obtained. To critics who over-simplified, it appeared on first analysis that certain aspects of the traditional theories of consolidation were in opposition to the technique.

In a single paper, it is impossible to present all the theoretical aspects of dynamic consolidation and therefore attention will be concentrated on the fundamentals of the technique best illustrated by an analysis of the complex behaviour of saturated alluvia during a tamping operation.

FUNDAMENTALS

As is often the case in the history of techniques, experience preceded theory, the need created the tool. Before having effectively carried out heavy tamping on saturated clay soil, the Authors thought it was impossible to consolidate it satisfactorily. However it was observed that these materials settled instantly several tens of centimetres and then discovered that they contained micro-bubbles of gas rendering them compressible under the effect of dynamic forces. The process of liquefaction appeared to us as a disturbing and deterring phenomenon when first apparent on alluvial soil. Later we noted that radial fissures appeared around the points of impact, and that these played a major part in the accelerated dissipation of pore-water pressures and even encouraged the appearance of geysers at the surface.

In order to comprehend the mechanism of dynamic consolidation, it was therefore necessary to carry out research on four main points which form the basis of its success:

(a) the compressibility of saturated soils due to the presence of micro-bubbles,
(b) the gradual liquefaction under repeated impacts,
(c) the changes of permeability of a soil mass due to the presence of fissures and/or the state of near liquefaction and the possible role played by adsorbed water,
(d) thixotropic recovery.

Each of these points will be considered in further detail.

Compressibility

It is customary to classify fine saturated soils as incompressible when subjected to rapid loadings, their low permeability opposing rapid drainage of pore-water. This evacuation of the water is considered to be a necessary and sufficient condition to allow settlement due to volume variations (theory of consolidation developed by Terzaghi).

However, early observations showed, surprisingly, that whatever the nature of soil treated, a tamping operation always resulted in an immediate considerable settlement; this result, under-standable for granular soils, could not be explained by traditional theories for saturated impermeable soils. Subsequent research showed that most quaternary soils contained gas in the form of micro-bubbles, the content varying between 1% for the most unfavourable cases and 4% in the more favourable.

As a first approximation, it could be assumed that the variations in volume of these micro-bubbles of gas may essentially be governed by the laws of Marriotte and Henry; in fact, other less known phenomena play a fundamental role. Thus, for example, shocks or mechanical vibrations modify the conditions of equilibrium of these micro-bubbles in a more or less irreversible manner—or at least with very pronounced hysteresis.

Liquefaction

As energy is applied to the soil under the form of repeated impacts, the gas gradually becomes compressed; as the percentage of gas by volume approaches zero, the soil starts to react as an incompressible material and at this stage, liquefaction of the soil begins to take place.   The energy level required to reach this stage is referred to as 'saturation energy'.   In practice, it should be noted that liquefaction in natural soils will often occur gradually; most natural deposits are layered and structured and the silty or sandy partings will liquefy before the clayey material.   It is most important that liquefaction of these layers or partings be obtained while the liquefaction of the main clay body must be avoided in order to prevent remoulding of the soil mass in the traditional sense of the word.   It is therefore imperative to know the precise level of energy corresponding to this threshold condition, which condition is essential to develop high pore-water pressures as well as reach high permeabilities.   The dynamic oedometer, a special instrument described later, was developed for that particular purpose.

It should also be noted that once the saturation energy has been reached, further application of energy would be entirely wasted apart from being damaging as already explained.

Permeability

A particular feature observed on dynamic consolidation sites was the initial very rapid dissipation of pore-water pressure which could not be explained by the coefficient of permeability measured before tamping.

When sands are subject to large hydraulic gradients (the conditions for piping), the permeability values are very high.   This phenomenon is actually general and apparent in all soils, whatever the granular size, when the conditions tend towards liquefaction.   A very slight local increase of pore-water pressure is sufficient to start a 'tearing of the solid tissues' by splitting, and quite naturally the flow of liquid concentrates in these newly created fissures.   This tendency to form fissures by splitting is very marked in natural soils, particularly if they have a pronounced macro-structure as a result of sedimentation of impurities; it is less so for soils that have been reworked or artificially homogenized (recent hydraulic fill).

By concentrating the tamping energy at regular grid locations, vertical fissures are created which are distributed regularly around the impact points.   These preferential drainage areas are generally perpendicular to the direction of lowest stress.   Founts of water which, under certain geological conditions, appear near the craters a few hours after tamping, are initiated and fed by this flow network.

On certain occasions, it has been noted that irregular and disorderly tamping can disrupt the continuity of these natural channels and render reconstitution more difficult for later and better planned tamping passes.

It has also been observed experimentally in the laboratory that the coefficient of permeability (and, in consequence, the consolidation coefficient) increases when the intergranular stresses decrease and that it reaches maximum value when the soil becomes liquid, at which instant the pore-water pressure is equal to the total pressure, γh.   This is partly why, during a dynamic consolidation operation which generally results in local conditions approaching liquefaction, high permeabilities can be observed and these are associated initially with very high pore-water pressures.

Finally, it would appear that the shock waves transform the adsorbed (solid) water into free water, thus encouraging an increase in the sectional area of the capillary channels; the reverse phenomenon occurs when the soil 'resets' under the influence of thixotropic phenomena. It should be pointed out that this information can only be proffered as a working hypothesis in so far as it is impracticable to measure the variations in the thickness of the layers of adsorbed water.

Thixotropic recovery

During a tamping operation, a considerable fall in shear strength is first noted, the minimum being recorded when the soil is liquefied or approaching liquefaction. At that time, the body of material is completely torn and the adsorbed water which plays an important part in stiffening of the structure is partially transformed into free water. As pore-water pressure dissipates, a large increase in shear strength and deformation modulus is noted; this is due to the closer contact between the particles as well as the gradual fixation of new layers of adsorbed water.

This latter process may continue for several months. This phenomenon, thixotropy, well-known in the case of sensitive clays, is in fact apparent in all fine soils (clays, alluvia, alluvial sands).

Graphical presentation

The foregoing fundamental points may be summarized and the consolidation or actual behaviour of the soil explained by using a modified presentation of the well known hydraulic system of a cylinder filled with incompressible fluid and supported by a spring. Fig. 1 shows the two systems in parallel; they are differentiated by four main features as follows.

The pore-water filling the cylinder is considered as partially compressible due to the presence of micro-bubbles.

Friction exists between the piston transmitting the forces due to the superimposed load and the containing cylinder. This results in hysteresis in the interaction between the hydraulic pressure increase and the intensity of piston surcharge. From this it can be deduced that a pressure reduction in the liquid does not automatically result in piston movement or a change in the spring. This point illustrates a fact often observed in foundation soils: the diminution of pore-water pressure without a corresponding settlement of the construction being investigated.

The stiffness of the spring (a representation of the compression modulus of the solid framework) is generally considered as constant, a notion which is often invalidated by experience; in fact, considerable modifications of the compression modulus can be observed under the influence of alternating loading. The adsorbed water plays an essential part in this process; under the influence of fortuitous energy additions (increased temperature, vibrations and so on) it becomes partially free. This results in a weakening of the mechanical bond between solid particles which reduces the overall strength of the material.

Permeability, in the case of dynamic consolidation, is represented by a nozzle of variable section for reasons which have been explained previously.

The various stages of dynamic consolidation may be summarized by a series of graphs.

Figure 2 relates to the changes in the soil after a single pass. Curve 1 shows the energy applied to the soil by a series of impacts on the same spot, curve 2 the corresponding volume variation of the soil, curve 3 the corresponding evolution of pore-water pressure in relation to the liquefaction pressure and curve 4 the evolution of the bearing capacity as a function of time.

Figure 3 relates to the same parameters as Fig. 2 but for a series of passes. It should be noted that although the energy follows an arithmetic progression, the volume changes and bearing capacity do not follow the same law.

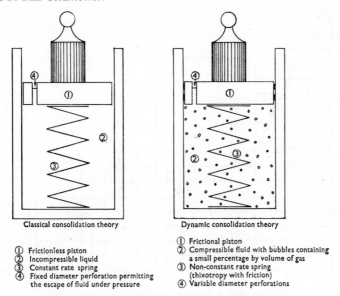

Classical consolidation theory

① Frictionless piston
② Incompressible liquid
③ Constant rate spring
④ Fixed diameter perforation permitting
   the escape of fluid under pressure

Dynamic consolidation theory

① Frictional piston
② Compressible fluid with bubbles containing
   a small percentage by volume of gas
③ Non-constant rate spring
   (thixotropy with friction)
④ Variable diameter perforations

Fig. 1.   Comparison of traditional and new theories of consolidation

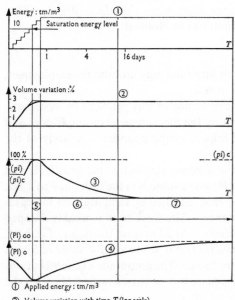

① Applied energy : tm/m³

② Volume variation with time T (log scale)

③ Ratio of pore-pressure PI to liquefaction
   pressure (pi) c against time T

④ Variation of bearing capacity of ground
   with time T

⑤ Liquefaction phase

⑥ Pore-water pressure dissipation phase

⑦ Thixotropic phase

Fig. 2.   Change in the soil after consolidation phase

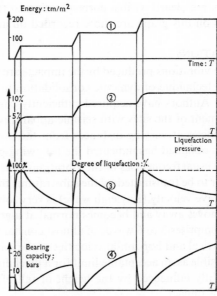

① Applied energy in tm/m²

② Volume variation against time (normal scale)

③ Ratio of pore-pressure pi to liquefaction
   pressure pi (c)

④ Variation of bearing capacity

Time between passes varies from one to four weeks according to the soil type

Fig. 3.   Variation of a soil subjected to series of dynamic consolidation passes

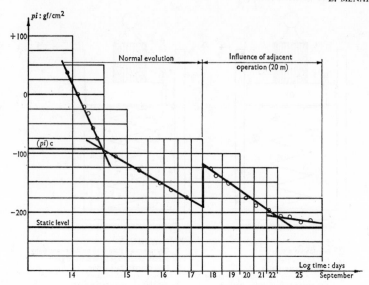

Fig. 4.   Dissipation of pore-pressure after a dynamic consolidation phase (Botlek)

Figure 4 shows the dissipation of pore-water pressure as a function of time.   Two distinct phases are clearly visible corresponding to the various phenomena explained already.   The values on this graph are those recorded on an alluvium site in Holland.

VIBRATIONS

The vibrations produced by the impacts are relatively large and may prohibit the employment of the technique of dynamic consolidation in urban areas.

The Authors have acquired sufficiently complete experience of this problem by systematic equipment of the sites with seismic apparatus permitting the measurement of amplitudes and frequencies at varying distances from the points of impact of the pounder.   A study of the results obtained has indicated the following points.

The usual frequencies of the vibrations caused by the tamping vary between 2 and 12 Hz and appear to be transmitted by the substratum; the most frequent value is in the order of 3 to 4 Hz. The wave velocity (Rayleigh wave) is very low in the zone liquefied by tamping, and increases as it moves away and becomes normal at a great distance.   The wave train is weakly damped and comprises 3 to 6 waves of almost constant amplitude.   At 30 m from the point of impact, the vertical and horizontal velocities of the soil particles remain much below the value of 5 cm/s admissible as an acceptable limit for a dwelling construction.   The amplitude of the vibrations is slightly influenced by the height of fall of the pounder, but increases noticeably with the area of impact.

CONTROL AND ENGINEERING OF A DYNAMIC CONSOLIDATION OPERATION

Before a site can be considered for a dynamic consolidation operation, it is necessary to carry out a soil investigation which should include

   (a) in situ testing such as pressuremeter, vane and penetrometer tests,
   (b) sufficient samples to carry out moisture content, Atterberg limits and particle size
        distribution determinations,

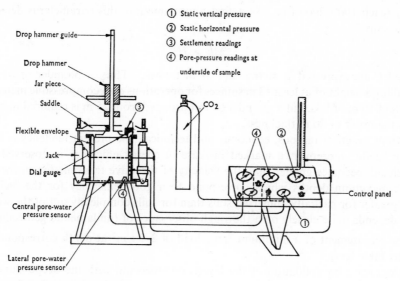

Fig. 5. **Dynamic oedometer (patented)**

(*c*) sufficient undisturbed samples for visual examination by splitting and testing in the cyclic and dynamic oedometers,

(*d*) sufficient boreholes to provide a stratigraphic description representative of the site.

Before engaging in a dynamic consolidation operation, it is important to predict, with some precision, the improvement that may be expected in the foundation soil; it is also appropriate to establish the operating method and the parameters for carrying it out: instantaneous settlement, saturation energy (per phase), number of phases, total energy and so on. When making these estimations, some reliance is placed on the results from equipment developed with this end in view, such as the dynamic oedometer.

The apparatus, such as presently in use (Fig. 5), is in the composite form of a triaxial apparatus and an oedometer, permitting the successive static consolidation of a 30 cm diameter sample, simultaneously transmitting static and dynamic loads to the sample and measuring, as a function of time, pore-water pressure, horizontal pressure and the corresponding settlements. The test which now has been standardized, permits determination of the saturation energy and the number of phases necessary to obtain a certain densification, the time for the dissipation of pore-water pressure and therefore the delay between phases, the predictable settlement under the action of tamping and the variation in the shear strength by means of a laboratory vane introduced into the sample through the upper piston.

Although it may be difficult in the laboratory to simulate the real behaviour of the soil, the results so obtained have permitted fairly accurate determination of evolution of the soil under the influence of tamping.

It is often necessary to carry out a few dynamic oedometer tests for a single site, in particular when the soil investigation has revealed layers of soils of a different nature. The lack of sufficient testing at this stage could lead to an erroneous design in the tamping pattern and energy requirements. The choice of pounder weight M and the falling distance h depend in the first instance on the thickness H of the layer to be compacted. The energy per blow Mh is an essential parameter; it varies in current practice from 150 tm to 500 tm, but in exceptional cases may reach 1000–2000 tm. As a first approximation, this parameter is determined by the relationship

$$Mh > H^2,$$

when h and H are expressed in metres and M in tonnes. Thus, a pounder of 8 t minimum weight, falling a height of at least 13 m suffices for operations on a compressible material having a thickness of 10 m or less, but a pounder of 16 t dropped from 25 m is required to consolidate a layer in the order of 20 m thickness.

For various practical reasons, economic and technical, the present tendency is to increase the falling height; moreover, a marked increase in energy efficiency is observed when the impact velocity becomes greater than the velocity of the wave transmission in the liquefying soil. The maximum depths which correspond to a good efficiency for the technique are distinctly greater for partially immersed soils than for soils completely out of the water. The efficiency depends quite closely on a thorough observance of fundamental geotechnical rules

(a) the requirement of a minimum static load of 2–3 t/m² at depth corresponding to the water table level,

(b) progressive consolidation of the layers, commencing with the deepest and finishing with the surface, by means of an adequate distribution of impacts.

Finally, the efficiency is a function of the shape and dimensions of the pounder, the height of its fall and the periods of delay observed between phases.

When a site of dimensions $L \times W$ is compacted uniformly over the whole surface, a peripheral fringe appears, with characteristics which are intermediate between the exterior non-compacted and the interior compacted zones. This fringe follows the line of the perimeter and has a width equal to approximately $2H$, when H is the thickness of the layer being consolidated. The uniformly improved zone then has the dimensions $(W-2H)$ and $(L-2H)$. It is therefore necessary to provide the extra width relative to the area to be effectively prepared. In particular cases, the width of this fringe may certainly be reduced by, for example, increasing the energy applied at the periphery, but this involves complementary inspections to avoid the appearance of 'frontiers' in the densification of the soil with consequent differential settlements beneath the construction.

SITE REQUIREMENTS AND CONTROL

The site to be consolidated must first be prepared, if only to support the weight of the tamping machine (60–200 t), and must be safeguarded against bad weather if sensitive to rain water (alluvia and clays) and removal of water rising to the surface during the consolidation process must be facilitated by means of peripheral trenches, drains, and so on.

Under the influence of dynamic consolidation, some water rises to the surface and inundates the low points; its removal may be expedited by pumping out the craters. In certain cases (clay soils saturated to the surface) it is useful, as a preliminary, to arrange horizontal drains generally constructed by trenching to a depth of 2–3 m and filling in with sand and gravel with perforated plastic pipes at the bottom of the trench.

The choice and the method of carrying out these auxiliary operations are important factors as the speed of the procedure of consolidation depends on them, especially during the winter months.

The control during the operation is carried out with

(a) penetrometer and pressuremeter for the measurement of strength and compressibility; in fine-grained soils, these tests are greatly influenced by the delay at which they are carried out after a pass; for final control, a minimum delay of 3–4 weeks should be observed,

(b) numerous piezometers placed at different elevations to determine the minimum delay between each pass,

(c) gamma densitometer and water content measurement of samples to check on the variation of dry density layer by layer,

(d) topographical measurements of the ground surface for determination of the overall variations of dry density.

SETTLEMENTS

The reduction of the settlements obtained due to dynamic consolidation is yet more distinct than the increase of the bearing capacity. The tamping produces a true pre-settlement w_c of the soils, well beyond the settlement w_o which would have occurred as a result of the weight of the construction only, without any preliminary consolidation. The ratio w_c/w_o measured after dynamic consolidation, varies between 2 and 3 compared with the values 0·8 or 0·9 usually obtained in the case of a traditional static preloading.

A particular point which should be mentioned concerns the secondary settlements which, by definition, appear at the end of the dissipation of pore-water pressure; as far as this Paper is concerned, no theory permits a prediction of the real long-term behaviour of a treated soil. The results to date are very favourable without being able to measure and explain them completely. It would appear that the secondary settlements are reduced in the same proportion as the primary settlements. To explain this result, an analogy may be applied: the secondary settlement is essentially due to the delayed deformation of the solid body of material under the influence of internal stresses; a metal structure built of numerous welded elements is itself subject to considerable internal stresses. These stresses can be eliminated by raising the temperature of the metallic part to a high value and allowing it to cool uniformly and slowly. It is similar in the case of soils: the liquefaction induced by dynamic consolidation results in greatly reducing, if not eliminating, the internal stresses; the body of material then sets with this new more stable equilibrium due to the part played by the adsorbed water which solidifies after reaching this new internal equilibrium. The settlement produced by the internal stresses is, in consequence, very reduced. Recent research, carried out in this particular field, has produced particularly interesting results, as much from the theoretical as the practical viewpoint, with regard to foundations.

CORRELATION WITH FIELD RESULTS

Karlstad (Sweden)

A stocking zone of 110 000 m², with a loading and unloading quay, was reclaimed from Lake Vänern by means of hydraulic fill with sandy silt material to a thickness of 2–10 m (Hansbo et al., 1973). The fill rests on 10–20 m of river silt deposits and clayey sands, underlaid at greater depth by glacial clays (Fig. 6). The strength characteristics represented [by two in situ tests indicate layers with poor bearing capacities and very variable compressibility (Figs 7 and 8).

The project included 3 sheds of 20 000 m² each, open air stocking areas, an office block and a sheet pile quay of 10 m freeboard. The aim was to obtain, by means of dynamic consolidation, an allowable bearing pressure of 3 bars at the surface and to reduce the differential settlements at least to 1 cm over 10 m under a uniformly distributed stocking load of 10 t/m².

The site was dynamically consolidated with a mean energy of 240 tm/m². The operation, carried out between November 1972 and April 1973, was not hindered by the low winter

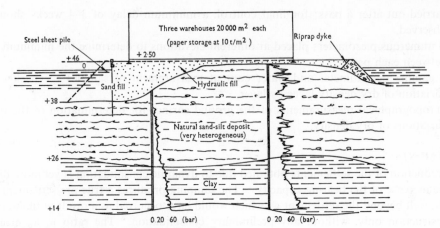

Fig. 6.   Section of installations at Karlstad Terminal

temperatures.   The effectiveness of the consolidation was checked by more than 250 static penetrometer tests and 72 pressuremeter profiles.

The improvement in the strength parameters of the soil is very evident in Figs 7 and 8.

The deformation moduli (measured by pressuremeter) increased by more than 500% for the first 7 m and more than 200% at 15 m depth.   The increase of the point resistances measured with the static penetrometer was itself appreciable, whereas before the dynamic consolidation operation it varied between 10 and 60 bars; at the end of the work, the mean values measured were more than 150 bars at 6 m and above 50 bars at 10 m depth.

The initial heterogeneity, incompatible with the project specifications, has been greatly reduced and the calculations derived from the geotechnical investigation results predicted a differential settlement below 1/1000.   The retention of an allowable bearing pressure of 3 bars for the foundation footings has been possible.   Erection of the structures was completed in the autumn of 1973.

Teesside (England)

The site lies in an area of reclaimed land to the northeast of Teesside in Cleveland.   It was until recently an estuarine mud flat close to the mouth of the River Tees.   The area has been reclaimed by infilling slag bunds with materials obtained by dredging operations in the river to form tanker berthing facilities.

Areas totalling approximately 50 000 m² were treated with a view to erecting oil storage tanks.   At the time of writing this Paper, construction was nearing completion.   The whole site was covered by an average of 4 m of made ground consisting of dredged sands and gravels with some silt; the compressibility of material was very variable; immediately below this hydraulic fill is a layer of black organic silty clay varying in thickness from 0·3 to 0·75 m. This overlies a natural deposit of medium fine single size sand which has a thickness varying from 9 to 17 m.

The site was treated with an energy level varying between 200 and 400 tm/m² depending on the conditions encountered; as an example, under one particular tank one side had to be treated with more than double the energy of the other side in order to obtain homogeneous conditions; the initial heterogeneity is also shown by variation of the enforced settlements between 25 and 45 cm.

The average improvement of settlement characteristics, reflected by the harmonic mean of the modulus of deformation, was in the order of 200–300%.

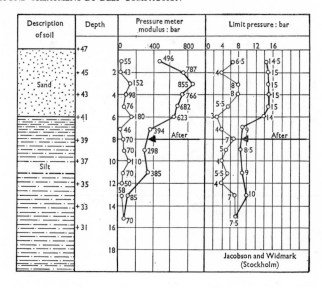

Fig. 7. Comparison of results obtained with pressuremeter: improvement of pressuremeter characteristics of soil—mean values for 20 borings made on the site

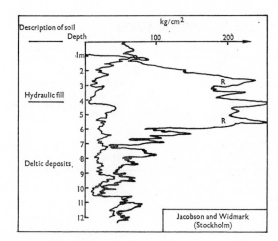

Fig. 8. Comparison of the results obtained with penetrometer: typical results obtained from static penetrometer boreholes—increase of point resistance in sandy silt is due to dynamic consolidation

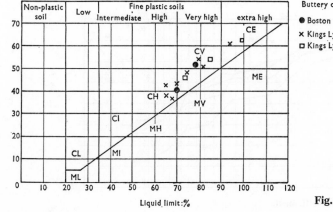

Fig. 9. Classification A-line plot

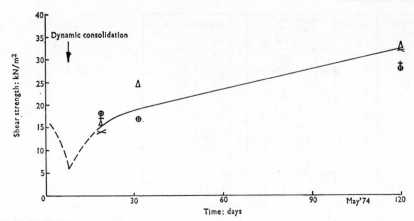

Fig. 10.   Shear strength

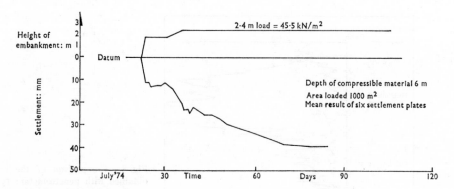

Fig. 11.   Test embankment—settlement

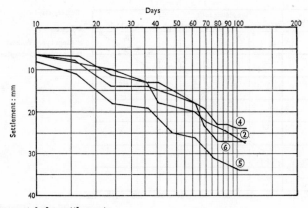

Fig. 12.   Eben–Tauern autobahn settlements

Fig. 13

An interesting point, indicating the depth of penetration of the treatment, was the 1 m rise in pore-water pressure recorded by a piezometer placed at a depth of 19 m in a soft alluvium layer encountered in one part of the site.

Boston Sewage Disposal Works (England)

The site lies three miles southeast of Boston near the village of Fishtoft in Lincolnshire.

Approximately 35 000 m² were treated by dynamic consolidation in order to improve the soil conditions for the erection of a new sewage disposal works. Binnie & Partners, Consulting Engineers acting on behalf of the owner, the Anglian Water Authority, carried out the settlement measurements on the surcharge area.

Approximately 25% of the site consisted of up to 6 m of non-stratified soft plastic clay known as 'buttery' clay; this clay has been described at length by Skempton (1945) and Wilkes (1972). Typical characteristics are

$$LL = 68-75\%$$
$$PL = 25-28\%$$
$$m_o = 50-100$$
$$\gamma = 1 \cdot 60 \text{ t/m}^3$$
$$c_u = 12-20 \text{ kN/m}^2.$$

Figure 9 compares the clay encountered at the Boston site with the sites investigated by Skempton (1945) and Wilkes (1972).

Figure 10 shows the variation of shear strength with time. The strength was recorded by field vane tests carried out at 17, 31 and 118 days after the last pass in the dynamic consolidation process. All the test results shown were taken within a radius of 1·5 m of the original borehole.

Calculations based on pressuremeter tests results indicated a total maximum settlement of approximately 49 mm under the load of 45 kN/m² applied by 40 m diameter filter beds. In order to check the validity of these computations, it was decided to surcharge an area of 1000 m² with a 2·4 m embankment. Figure 11 shows the settlement records which, after two and a half months, appear to have stopped at 40 mm. It is interesting to note that a test embankment carried out by Wilkes (1972) on similar ground yielded 150 mm immediate settlement (after four days) with no sign of stabilization.

Eben (Austria)

The site is in a valley in the Austrian Alps; a motorway interchange had to be built over compressible layers of organic silts and peats. The compressible layers varied between 6 and 14 m in depth; the water table was near the surface and the average water content varied between 150 and 400% and in certain areas reached 1200%. Approximately 80 000 m² of the site was treated by dynamic consolidation with a fairly high energy level in the order of 600 t/m². A 2·5 m thick layer of fill (glacial silty sand and gravel) was placed over the embankment areas; up to seven passes were performed; additional fill was brought in after each pass to make up for the enforced settlement and maintain the working platform at 2·5 m above ground level. Lateral trench drains were installed alongside the embankments to facilitate the evacuation of water. The embankments were then lifted to 7 m above ground level and subjected to a light treatment of dynamic consolidation to compact the 4·5 m freshly brought on. Settlement plates were then installed on the crown of the embankments. Typical settlement results are shown in Fig. 12 indicating an average value of 30 mm after 113 days.

CONCLUSION

The apparent simplicity of dynamic consolidation on the site may often obscur the complexity in the realm of soil mechanics and the development of new equipment to fit the technical requirements. Extreme complexity from the viewpoint of soil mechanics, which sometimes deterred the best minds in this field, necessitates a high degree of competence in the control of a dynamic consolidation operation, as well as substantial previous experience. Systematic study on site and in the laboratories of the different phenomena encountered have allowed the technique to be refined, thus improving efficiency; the fundamental research undertaken has opened new horizons in the understanding of the behaviour of soils.

The theoretical understanding of the technique soon made obvious the imperative necessity to develop new equipment for the laboratory as well as the field; a striking example is the machine shown in Fig. 13 capable of dropping a 40 tonne tamper from a height of 40 m.

Beyond its application in the classic realm of construction, new perspectives appear on the horizon: earth dams, artificial islands for atomic power stations, oil storage or airports.

One of the most important facets of dynamic consolidation is perhaps that soil can now be considered as a material which can be engineered to technical requirements.

ACKNOWLEDGEMENTS

The Authors wish to thank the Lincolnshire Sewage Division of the Anglian Water Authority for permission to include data obtained on the Boston site.

REFERENCES

Hansbo, S., Pramborg, B. & Nordin, P. O. (1973). The Vänern Terminal; an illustrative example of dynamic consolidation of hydraulically placed fill of organic silt and sand. *Sols-Soils*, No. 25.
Skempton, A. W. (1945). A slip in the west bank of Eau Brink cut. *Proc. Instn Civ. Engrs*, May, 267–287.
Wilkes, P. F. (1972). Kings Lynn trial embankment. *Jnl Institute of Highway Engineers*, Aug.

Soil compaction by impact

R. A. SCOTT* and R. W. PEARCE†

The problems of prediction of the reaction of real soils to intense impact are discussed and the difficulties enumerated. Idealized models are presented to illustrate the parts played by stress wave reaction and by inertia forces associated with elasto-plastic void closure in determining the deceleration of a falling weight and the utilization of the impact energy. The general behaviour of real soils to a specified impact is then reviewed by selecting more specific types of soil and discussing their expected behaviour in the light of the general principles embodied in the idealized models.

Les problèmes de prévision de la réaction des sols réels au contact d'un choc intense, sont discutés et les difficultés énumérées. Des modèles idéalisés sont présentés afin d'expliquer les rôles joués par la réaction d'ondes de contrainte et des forces d'inertie associées à la fermeture de la cavité élasto-plastique dans la détermination de la décélération d'un poids tombant et l'utilisation de l'énergie de choc. Le comportement des sols réels à la suite d'un choc particulier est alors examiné en sélectionnant des types de sols spécifiques et en discutant le comportement prévu compte-tenu des principes généraux établis dans les modèles idéalisés.

The densification of soils by moderate steady load has received much attention, emphasis being primarily on the time effects involved in the displacement of excess water as the soil skeleton condenses. Much less consideration appears to have been given to densification under impact loading, except at the practical levels.

Tamping devices are widely used for the compaction of unsaturated soils. Several mechanized forms of falling weight compactor are in common use and Lewis and his colleagues at the Transport and Road Research Laboratory have tested a number of such machines and have found, for example, that compaction depth depends very markedly on the impact energy available per unit area of contact (Lewis, 1957; Parsons and Toombs, 1968). In recent years Ménard has shown that impact devices with extremely high energy of impact can compact to depths of many times the diameter of a wide impacting mass and have a significant densifying action on certain soils in their saturated state.

Some general indication of the practical performance of falling weight compactors is given in Table 1. The first three entries relate to the moderately intense machines tested at TRRL and the final entry to the partially saturated silty sand of hydraulically placed fill at Teesside, compacted by an intense Ménard machine.

Similar and greater depths of compaction have been recorded by Ménard (1972, 1974) on other soils.

* Cementation Specialist Holdings Limited.
† Cementation Ground Engineering Limited.

This Paper is presented in an attempt to provide a broad descriptive picture of compaction by impact. The principal mechanisms which act to retard a falling weight on contact with the surface of a medium are exemplified for a range of idealized soil models and consideration is then given to impact on real soils.

NOTATION

a	contact surface radius	V	impingement velocity of falling weight
c	dilation velocity in soil	V_1	initial velocity of common contact
c_w	dilation velocity in weight	v	particle velocity in radiated stress wave
E	Young's modulus	z	instantaneous position of compaction front
h	final compaction zone depth		
k	$\rho_c/(\rho_c - \rho)$	ν	Poisson's ratio
M	mass of weight	ρ	soil density
m	mass per unit area of weight	ρ_c	compacted soil density
R	damping constant	ρ_w	density of weight
S	spring rate constant	σ	vertical stress at soil contact (general)
t	time after impact	σ_0	initial vertical stress at soil contact
u	soil contact deflexion	σ_L	idealized limiting elastic stress of soil
$\dot{u}$	soil contact velocity	ω	angular resonant frequency
$\ddot{u}$	soil contact acceleration		

IMPACT OF A FALLING WEIGHT

Whatever may be the detailed deformational properties of a medium, elastic or otherwise, the immediate consequence of a surface impact is a highly localized contact stress which originates in momentum sharing between the thin slices of the respective materials confronting the contact surface. The initial stress level and therefore the deceleration of the weight depends on dynamic properties of the media such as the specific acoustic wave impedance. Subsequent behaviour such as decay of surface stress, pulse duration and extent of penetration of the main movement are determined by the way in which the body of the impacted medium reacts to the movement imposed at the surface. For example, the interaction is widely different according to whether the main restraint of the indenting movement at the surface is bulk compressibility, shear restraint from the medium flanking the indenting zone, or inertial forces accompanying either easy collapse of voids or gross heaving motion to the nearby surface.

Soils are likely to fit variously into this range of possibilities. It might be expected that dry and well compacted soil would behave approximately as in classical elastic impact and that very loose dry soil would yield to the motion of the falling weight by a progressive and energy

Table 1. Typical performance of falling weight compactors

	Base area, m^2	Falling mass, kg	Height of fall, m	Energy per unit area, J/m^2	Impact velocity, m/s	Recorded peak soil stress, kN/m^2	Approximate consolidation depth for two passes, m
RRL (1)	0·21	174	2·46	20 000	4·9		0·2
RRL (2)	0·09	206	0·61	13 500	3·4	630*	
Arrow D500	0·09	588	2·24	143 000	6·6		0·4–0·6
Ménard	6·0	25 000	18	735 000	18·8	725†	8

* 150 mm below contact.
† 2450 mm below contact.

absorbing compactive collapse. The reaction of saturated soils is less easily estimated as high transient pore-water pressures would moderate or destroy the rigidity of the soil matrix and allow a possibility of a gross, impulsive plastic deformation extending to the ground surface. The nature of these various restraint mechanisms will be illustrated in the course of this present section of the Paper by consideration of simple idealized soil models. More realistic descriptions of the performance of actual soils will be given later.

Ideal elastic unsaturated soils

While perfectly elastic soils cannot by definition show compaction behaviour, a very wide range of unsaturated soils, including loose soils, shows elastic effects as illustrated, for example, by capacity to propagate seismic waves with only modest attentuation. According to the physical data recorded in Clark (1966), the dilational velocity of unsaturated sands ranges from 200 to 1000 m/s with the lower values relating to loose conditions. The ratio of dilation to shear velocity is typically about 2·5 and this corresponds to a Poisson's ratio of 0·405. Practical data are therefore available for inclusion in models of dynamic behaviour.

The main features of the impact of a falling weight on the surface of an 'elastic soil' can be derived from established mechanical principles (Kolsky, 1963; Lysmer and Richart, 1966). At the instant of impact the velocity of the lowest surface of the weight falls from the impingement velocity V to a new value V_1 which, because of contact, is also the initial displacement velocity of the uppermost soil layer beneath the weight. Impact is accompanied by a localized zone of very high stress and in consequence, step-fronted stress waves are transmitted at the appropriate seismic velocity out into the body of each medium. These stress waves preserve the conservation of momentum and transmit the general sudden local velocity changes further into the respective media. The seismic waves commence as plane dilation waves and it is generally well known (Timoshenko, 1951) that the common stress level in these transmitted stress waves is given by the impulsive velocity change multiplied by the product (acoustic impedance) of density ρ_w, ρ and dilation velocity c_w, c of the respective media of weight and soil. For the soil medium the initial stress level is therefore $\rho c V_1$; for the weight medium the initial stress is $\rho_w c_w (V - V_1)$ and these stresses must clearly be equal. The weight will be regarded as a shallow circular prism of dense and rigid material. The stress wave transmitted into the weight is therefore reflected at the upper surface of the weight and returns as a tension wave to interact further with the common contact stress. The surface is thereby made to change its velocity in reiterated jumps. However, the time of double-transit of waves in the weight is small (about 1 ms) and the velocity change $(V - V_1)$ is considerably less than V because of the general disparity of acoustic impedance between the dense weight and the soft soil; in consequence the mean speed of the weight never differs greatly from the velocity of the common contact surface and it is sufficient and convenient for present purposes to regard the weight as a 'lumped mass' with a continuously changing velocity which has an initial value equal to the full impact velocity. The initial stress in the soil is therefore

$$\sigma_0 = \rho c V \qquad . \quad . \quad . \quad . \quad . \quad . \quad . \quad . \quad . \quad (1)$$

The later stress history of the pulse is set by the cumulative effect at the contact zone of stress waves which while continuously radiated down from the contact surface are also partially reflected back from the soil to build up further stress at the surface. The reflexion effects are complex and arise, for example, from coupling of the initial dilation wave to shear waves and to Rayleigh surface waves. It is unnecessary to follow these waves in detail as their total effect is better expressed in the results of the analysis of the problem by writers in the field (Miller and Pursey, 1954; Bycroft, 1956). These authors show that the surface stress level σ

is related to the average deflexion u and velocity $\dot{u}$ of the contact surface of radius a by a relation approximating over a wide span of frequency and elastic moduli to

$$\pi a^2 \sigma = R\dot{u} + Su \qquad \ldots \ldots \ldots \ldots \quad (2)$$

where R is a damping constant and S is a spring constant closely equivalent to that deduced by the Boussinesq method for steady loading

$$S = 2aE/(1-v^2) \qquad \ldots \ldots \ldots \ldots \quad (3)$$

The coefficient of the resistive term approximates at all but very short intervals after impact to

$$R = 0.6\,\pi a^2 \rho c \qquad \ldots \ldots \ldots \ldots \quad (4)$$

The discrepancy of 0·6 between the initial surface stress deduced from equations (1) and (2) arises from different representations of the variation of pressure across the circle of contact and does not seriously affect the main argument.

The relation in equation (2) ceases to hold for very high frequency components of the surface motion for which the characterizing stress wave wavelength is small compared with the contact circle diameter; however, this limitation affects the pulse shape only in the first few milliseconds after impact.

The equation for the deceleration of the weight of lumped mass M is

$$-M\ddot{u} = R\dot{u} + Su \qquad \ldots \ldots \ldots \ldots \quad (5)$$

whence the deflexion u and the mean stress σ at the soil contact are given by

$$u = \frac{V}{\omega}e^{-(R/2M)t}\sin \omega t \qquad \ldots \ldots \ldots \ldots \quad (6)$$

and

$$\sigma = \frac{SV}{\pi a^2 \omega}e^{-(R/2M)t}\cos\left(\omega t - \cos^{-1}\frac{R\omega}{S}\right) \qquad \ldots \ldots \ldots \quad (7)$$

where ω^2 is $(S/M - R^2/4M^2)$. Clearly therefore ω is equal to angular frequency of free resonance of the total mass of the weight as restrained by the Boussinesq spring reaction of the soil at the contact surface and modified by the dissipation resistance R originating from energy absorbed by radiated stress waves.

It is apparent from equations (6) and (7) that when $R/2M$ is small (e.g. where M is large) the surface motion is periodic in form but with an overriding exponential damping. This form of surface deflexion is shown in curve B and the corresponding surface stress in curve A of Fig. 1. The soil parameters used for this pair of curves are set out in Table 2 and the soil has much the same dilational and shear velocities as would correspond to a compact dry silty sand. The mass, base area and impact velocity correspond to the final entry of Table 1 and therefore relate notionally to an intense Ménard dropping weight installation. The surface stress rises

Table 2. Assigned elasticities, densities and elastic limits

Model	Figure number	Curves	Initial density ρ, kg/m³	Compacted density ρ_c, kg/m³	Young's modulus E, N/m²	Dilation velocity c, m/s	Elastic limit σ_L, N/m²	Notional equivalent
Elastic	1 1	A and B C and D	1900 2000		10^7 $1{\cdot}4\times10^7$	116 1600	— —	Unsaturated silty sand Saturated silty sand
Elasto-plastic	4	A and B	1990	2030	—	1000	3×10^5	Uncompacted partially saturated silty sand

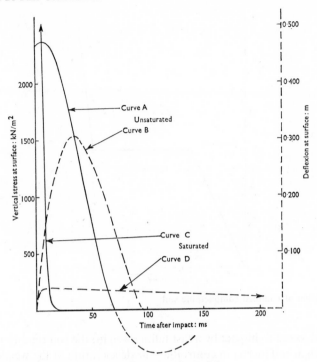

Fig. 1.   **Stress and movement at the impact surface, elastic model**

instantly to a moderately high level, namely 2350 kN/m^2 and remains in that region for 20 ms before dropping sharply to zero after 70 ms.   At this time the soil surface and the weight have passed through a point of maximum deflexion and are moving upwards.   The stress in the surface would be negative at times a little larger than 70 ms.   The weight in fact rebounds at this time and retains as kinetic energy about one tenth of the impact energy.   The remaining nine tenths of the impact energy have been transformed first into plane descending dilational waves and thereafter into a combination of dilational and shear waves.   Indeed most of this energy is carried away from the impact zone by Rayleigh waves concentrated near the ground surface.

The analysis for ideal linear elasticity requires only a small modification to account for any viscous resistance inherent in the soil, as for example where the restraining forces include an element proportional to rate of strain.   The parameter R must be given a suitably increased value.   The effect of viscous dissipation is to increase initial stress at the surface, accelerate decay and diminish, if not eliminate, rebound.   The impact energy is still radiated as stress waves, although the waves attenuate more severely during propagation.   Movements are still concentrated in the shallow zone lying within about one diameter of the contact surface as, for example, for steady surface loading.

Elasto-plastic unsaturated soil

Loose unsaturated soils subject to steady localized surface loading deform typically as shown by the curve A of Fig. 2.   The deformation is of a generally elastic nature at low stress levels and at these stresses the soils can propagate seismic waves.   With increasing stress the slope of the deformation curve falls more or less sharply due to the relative ease with which voids can be collapsed at the higher stress levels.

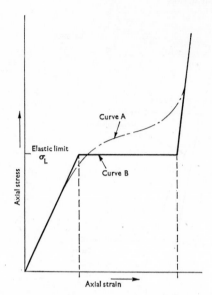

Fig. 2.   Axial deformation of confined compactable soil

If such a soil is subjected to impact by a fast falling weight, the soil rigidity may play a much less important role than soil inertia in controlling the deceleration of the weight and in absorbing the energy of the impact.   An idealized representation of a compactable soil in respect of these inertial and energy consuming effects in the elasto-plastic soil is represented by curve B of Fig. 2.   The stress level of the plateau has been chosen to lie in the region of the reduced slope.

A three-dimensional treatment of the reaction of the soil underlying the contact is impracticable as the strains are generally so large that the shear restraints due to flanking regions of soil are not easy to quantify.

However, when the impact momentum is high the weight will punch through the upper soil layers and carry down a growing zone of compacted material of a generally cylindrical shape. For present purposes of illustration we shall discount the inevitable lateral spread of the compacted zone and use a one-dimensional description based on the approach mapped out for example by Salvadori (1960).

Immediately upon impact the stress level rises because of stress wave reaction due to the elastic nature of the first small movements of the soil at the contact surface.   When the stress level has reached the level σ_L of the plateau, the soil particles at the surface have acquired a velocity v associated with a radiating stress wave which travels downwards into the medium with the seismic dilation velocity c appropriate to initial elasticity.   The wave is accompanied by a pressure front in which the axial stress is given by a form of equation (1) that is,

$$\sigma_L = \rho c v$$

The radiation of the stress wave is followed almost immediately by a further acceleration of the surface particles such as to bring the surface to the same instantaneous velocity V as the weight.

If z is the instantaneous position of the front of the steadily lengthening compacted material (Fig. 3) the retarding stress applied at the bottom surface of the weight is

$$-m\frac{\mathrm{d}}{\mathrm{d}t}(\dot{u}-v) = \rho_c \frac{\mathrm{d}}{\mathrm{d}t}\left[(z-u)\frac{\mathrm{d}}{\mathrm{d}t}(u-vt)\right]+\sigma_L \quad . \quad . \quad . \quad . \quad . \quad (8)$$

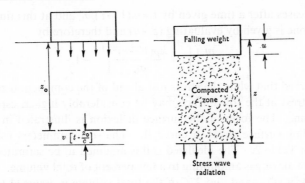

Fig. 3. **One-dimensional compaction**

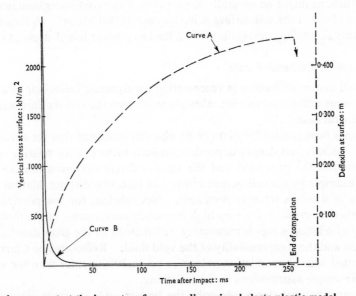

Fig. 4. **Stress and movement at the impact surface, one-dimensional elasto-plastic model**

where m is written for the ratio $M/\pi a^2$ and ρ_o is the compacted density.   The distances z and u can be shown to be related by the expression $z = k(u-vt) + vt$ where $k = \rho_o/(\rho_o - \rho)$.   This relation can be used to eliminate z in equation (8), with the result that

$$m\frac{d}{dt}(\dot{u}-v) + k\rho\frac{d}{dt}\left[(u-vt)\frac{d}{dt}(u-vt)\right] + \sigma_L = 0 \quad . \quad . \quad . \quad . \quad . \quad (9)$$

The displacement u of the surface is obtained by solving equation (9) hence

$$u = vt + m(F-1)/k\rho \quad . \quad . \quad . \quad . \quad . \quad . \quad . \quad . \quad (10)$$

where

$$F = \left(1 + \frac{2k\rho}{m}t - \frac{\sigma_L}{m^2}t^2\right)^{1/2}$$

The surface stress in the soil is then given by

$$\sigma = \frac{\sigma_L + k\rho(V-v)^2}{F^3} \quad . \quad . \quad . \quad . \quad . \quad . \quad . \quad . \quad (11)$$

Surface motion ceases after a time given by $t = m(V-v)/\sigma_L$ and at this time the final depth h of the compacted zone is given by evaluating $(z-u)$ and therefore by

$$h = \frac{m}{\rho_c}\left\{\left[1 + \frac{k\rho(V-v)^2}{\sigma_L}\right]^{1/2} - 1\right\} \quad . \quad . \quad . \quad . \quad . \quad . \quad (12)$$

It should be observed that while the stress just ahead of the compaction zone is at the elastic limit stress σ_L the stress at the soil surface may be considerably higher, especially at the early stages of compaction. The form of the surface deflexion is illustrated in curve A of Fig. 4 and the corresponding surface stress in curve B. The soil parameters used for this pair of curves are set out in Table 2. The idealized soil is assumed to be saturated with water apart from the occlusion of air or gas amounting to a few percent of total volume. The idealized soil notionally represents a silty sand and 2% of the total volume is assumed to be closed during compaction, through the collapse of gas-filled loose zones. The dilation velocity has been taken as 1000 m/s, as might be generally appropriate if the void water contained trapped gas. The initial stress level at the soil surface is in this case 35 000 kN/m². A larger proportionate change in density following compaction would lead to a lower initial stress at the surface.

Linear elastic soil in a saturated state

There are well known difficulties in representing the dynamic behaviour of a fully saturated elastic soil, as some differential motion takes place between the soil skeleton and the pore-fluid in the most general case.

The subject has been studied by Biot (1956) who demonstrated that the effect is to make the equivalents of elasticity and density dependent on such factors as the relative compressibilities of the skeleton and the pore-fluid and the viscous forces opposing relative movement, as expressed for example by the soil permeability. In fact, two distinct dilation waves are possible, only one of which is strongly persistent. Nevertheless, for low permeability soils and with some further reservations for very high frequency components of the motion, the propagation velocity of dilation waves is reasonably well defined by the gross density, the rigidity of the soil skeleton and the compressibility of the void fluid. References by Clark (1966) to field data for saturated soils show dilation velocities up to 20% greater than for water, as would follow from the simple assumptions in the foregoing.

It would seem a natural step to describe the impact on a compact saturated soil by the classical model embodied in equation (5), using parameters R and S that reflect the rigidity modulus of the dry soil matrix and the bulk compressibility of the pervading water. If the intergranular contacts were still retained under transient conditions following impact, this approach might be acceptable. The wave impedance R would increase considerably because of the higher dilation velocity and slightly increased density. The spring rate S would change little, as it chiefly depends on soil matrix rigidity.

Curves C and D of Fig. 1 illustrate the surface stress and surface movement for this model. The soil parameters are set out in Table 2, and the example is again evaluated as if the mass, base area and impact velocity correspond with the final entry of Table 1. The surface stress rises instantly to 36 000 kN/m² but has fallen to a very much lower value 10 ms after impact. There is no ground resonance; the parameter ω takes the imaginary form and the main cosine term of equation (7) gives place to a hyperbolic term. Elastic recovery proceeds slowly under the driving force provided by the soil rigidity.

This simple classical model takes no account of the effect of transient pore-water pressure on the interparticle contacts. Nevertheless, the exceptionally initial high stresses predicted originate in the bulk compressibility rather than the rigidity, and very high intensity stress waves can be expected for impact both on submerged soils and on saturated soils unless other modes of deformation such as impulsive heave intervene to limit the stress.

Saturated compactable soil

While modest levels of stress might be expected to cause collapse of hitherto loose zones in dry soil, saturation implies that such cavities in general are filled with pore-water. Collapse would then be greatly inhibited by resistance of the water to compression and a transient stress pulse would at most destroy some arching structures near a proportion of potential compaction zones and initiate some small amount of long-term consolidation under ambient ground stress.

However, real saturated soils may have a few percent of the total void space occupied by occluded air or gas. The stress wave intensity would certainly be reduced because of the lowering of the bulk modulus of the void fluid, but the impact stresses would still ensure that a loose zone or cavity would collapse provided the air content in the zone was high. In this event the stress limiting action of the closing voids, described earlier for the case of collapsing zones in a dry elasto-plastic soil, would modify the stress level accompanying the impact and, if the proportion of air-filled cavities was sufficiently high, would in fact control it.

The growth of the compacting zone, the duration and the depth of penetration of the soil movement would again be described by equations (10) to (12), but with the density parameters reflecting changes of packing of zones containing at least a little air.

DISCUSSION

The full solution of the problem of impact on a real soil obviously presents formidable analytical difficulties since the forces and motions are not likely to be representable by simple elastic or by simple plastic descriptions like those set out for the special conditions and particular geometries above.

However, some progress should be possible by searching for particular soil conditions that permit some sort of gross simplification and at least the suggestion of a qualitative answer. Historically at least a main application of tamping has been for the compaction of very loose, moderately dry soil. Impact velocities are quite low; hand tampers involve free fall velocities of about 3 m/s. If the soil were to react elastically—as it does approximately to light steady loading—the immediate impact stress would be in the region of 1200 kN/m² for a soil with a dilation velocity of, say, 200 m/s; for all except stiff soils, this stress would begin to fall into the inelastic régime. Nevertheless, for very light impacts, the ideal elastic model might be a rough general description and both stress and movement curves such as A and B of Fig. 1 might be approximately correct: a hand tamper might even rebound in some slight degree. A main part of the impact energy would be converted into seismic radiation, with an immediate highly localized plane dilation wave which then converts to a combination of dilation and shear waves compatible with the presence of the free ground surface.

For real tamping action, even at the hand tamper level, the higher stresses are certain to increase the effective damping constant, through the action of shearing forces or of energy loss from void closure. The medium and high intensity mechanical compactors produce impact velocities ranging up to 20 m/s and more, and it is therefore natural to expect inelastic behaviour in the soil where such machines are used.

The intermediate sort of impactor to which TRRL has devoted attention (Table 1) does not usually compact for a depth of more than one diameter, and the closure of voids is then very roughly confined to a short cylindrical prism below the contact area.

If these machines were operated with a wider base but with impact velocity and mass per unit area maintained, the compaction zone would be more nearly disc-like. The shear restraints acting from the soil flanking the disc would then be proportionally less important than the frontal forces and the one-dimensional elasto-plastic model might possibly give a recognizable description of the stress and movements in the soil and the development of the compaction zone.

In tentatively applying the one-dimensional conclusions to low and moderately intense impacts it is convenient to recast equation (12) in a form relating better to the utilization of the impact energy, thus if m is the mass per unit area of the base of the weight, and unit cross-section is alone considered, the appropriate form is

$$\sigma_{\mathrm{L}} h \frac{\rho_{\mathrm{c}} - \rho}{\rho} \left(1 + \frac{h\rho_{\mathrm{c}}}{2m}\right) = \tfrac{1}{2}m(V^2 - v^2) - mv(V - v) \quad . \quad . \quad . \quad . \quad . \quad (13)$$

The multiplying factor on the extreme left is the plastic closure energy in the compaction zone; the first term on the right is clearly the change in kinetic energy of the falling weight, and the final term is easily shown to be the total energy radiated in the plane dilational stress waves. If equation (13) is interpreted for low impact velocity on a poorly compacted, slightly cohesive soil such as a tipped clayey sand gravel the ratio $h\rho_{\mathrm{c}}/2m$ could be substantially less than unity and the plastic closure energy of the voids would be approximately equal to the available impact energy less the radiated strain energy. The impact energy is then used with reasonable efficiency. If, however, V and k are high, as might correspond to large drops and a moderately consolidated unsaturated silty sand, $h\rho_{\mathrm{c}}/2m$ becomes greater than unity and less closure is produced for a given amount of impact energy. This extra dissipation of energy is due to the impulsive nature of the entrainment of formerly uncompacted material into the compacted zone. During the process of collecting each additional thin zone of newly compacting material, the combined mass of weight and 'rigid' compacted zone reduce slightly in velocity so that axial momentum is conserved. However, this is an impact process with no rebound and no compensating forward emission of coherent stress waves to restore the energy balance. Some energy is therefore lost notionally in very short wavelength stress waves which are highly attentuated and short lived. In the dynamic sense appropriate to impact compaction of an elasto-plastic soil, the energy associated with closure is the full expression on the left-hand side of equation (13).

No more than a brief comment will be made on the many inadequacies of a one-dimensional theory of compaction. Shear reaction and strain coupling generally must lead to retardation and lateral spread of the compaction zone. The simple plane dilation stress wave is not viable for three dimensions. Coupling to shear and to secondary dilation waves would lead once more to partial reflexion of the primary wave back to the compaction front where it will in effect modify the relation between stress wave pressure σ_{L} and the associated particle velocity jump. Less energy would be radiated as stress waves and a little more energy would be made available for compaction.

The reaction of a fully or partially saturated soil to impact is a problem of considerable interest. A very low velocity impact on a saturated soil would lead, on the basis of an elastic model with fully retained particulate contact, to a relatively high but rapidly decaying stress wave. Nevertheless, the period of high stress would last a few milliseconds and give ample time for the stress wave to be propagated to the ground surface near the weight. This maintained stress gradient would then tend to produce some impulsive heave at the surface. In the case of real soils and high impact velocities, the high maintained gradient during the early history of the stress wave and the low or null interparticle forces caused by the high transient pore-water pressure greatly increase the tendency for surface heave and for direct lateral expulsion of soil from beneath the weight. In practice it is unusual to drop weights directly on to saturated or partly saturated soil. The standard procedure is to place about one metre of artificial surcharge and compact from a higher level. The stress wave is then somewhat attenuated by the time it reaches the saturated soil.

The surcharge has several independent actions. The total ground stress is increased and transient pore-water pressures have less tendency to destroy shear strength, at least in the outer parts of the slip field of the heave. The surcharge itself inhibits heave. Impact on the surcharge rather than on the saturated soil greatly reduces the impact stress and extends the duration of the pulse, much in the way that a pile 'dolly' acts.

Uncompacted saturated soils with impact applied to the surcharge would show only a very small volumetric strain and presumably a negligible direct collapse of loose zones. However, poorly compacted, partially saturated, soil with occluded air or gas would be subject to the same type of compaction action as exhibited by unsaturated soil. Compaction would be concentrated on the air-filled zones, but the accompanying shear movements near these zones might lead to some permanent destruction of local arching and ultimate long-term consolidation. The immediate densification would be confined to the neighbourhood of air-filled zones; very little strain would be recoverable.

Finally, it is interesting to reflect on the detailed reaction of an air-filled void to high transient stress. Certain requirements are needed to initiate the collapse of a hitherto stable void. For cohesive materials this would appear to be a stress field sufficiently intense to mobilize a plastic failure; for granular materials it may be sufficient to provide a shearing movement to destroy a critical arching action.

The extent of closure of a loose zone during impact will, however, depend on the availability of sufficient surplus energy, presumably principally local strain energy, to provide respectively: the elastic energy in the compressing occluded air; the kinetic energy needed to move surrounding material into the closing zone with sufficient speed; and the energy dissipated in the plastic shear of cohesive material or the viscous shear associated with the differential microscopic movements between particles and a pervading pore-fluid in granular materials. Each of these factors is therefore a source of delay in closure of loose zones and will be very dependent on soil material and on size and shape of cavities and voids. Certain sorts of open void are not likely to densify to the fullest extent during a single impact pulse.

ACKNOWLEDGEMENT

The Authors wish to thank Cementation Specialist Holdings Ltd for permission to publish this Paper.

REFERENCES

Biot, M. A. (1956). Theory of propagation of elastic waves in a fluid-saturated porous solid. *Jnl Acous. S.A.* **28**, 168–178.

Bycroft, G. N. (1956). Forced vibrations of a rigid circular plate on a semi-infinite elastic space and on an elastic stratum. *Phil. Trans. R. Soc.* A248, 327–368.

Clark, S. P. (1966). Handbook of physical constants. *Geol. Soc. Amer.* Memoir 97, Chapter 9, Table 9.5.

Forssblad, L. (1965). Investigations of soil compaction by vibration. *Acta Polytechnica Scandinavica C.E. and Bldg Constr.* Series No. 34, 85–111.

Kolsky, H. (1963). *Stress waves in solids.* New York: Dover.

Lewis, W. A. (1957). A study of some factors likely to affect the performance of impact compactors on soil. *Proc. 4th Int. Conf. Soil Mech.* **2**, 145–150.

Lysmer, J. & Richart, F. E. (1966). Dynamic response of footings to vertical loading. *Proc. Am. Soc. Civ. Engrs* **SM1**, 65–91.

Ménard, L. (1972). La consolidation dynamique des remblais récents et sols compressibles. *Travaux,* November.

Ménard, L. (1974). La consolidation dynamique des sols de fondation. *Conférences ITBTP.*

Miller, C. F. & Pursey, H. (1954).   The field and radiation impedance of mechanical radiators on the free
    surface of a semi-infinite isotropic solid.   *Proc. Roy. Soc.* A223, 521–541.
Parsons, A. W. & Toombs, A. F. (1968).   The performance of an Arrow D500 dropping-weight compactor
    in the compaction of soil.   *RRL Report* LR. 229.
Salvadori, M. G., Skalak, R. & Weidlinger, P. (1960).   Waves and shocks in rocking and dissipative media.
    Proc. Am. Soc. Civ. Engrs EM2, 77–105.
Timoshenko, S. & Goodier, J. N. (1951).   *Theory of elasticity*, Chapter 15.   New York: Dover.

A field trial of the reinforcing effect of a stone column in soil

J. M. O. HUGHES*, N. J. WITHERS† and D. A. GREENWOOD‡

The load–settlement relationship for plate loading of an isolated stone column in soft clay was predicted prior to field testing. The column was constructed by vibro-replacement and, after the test, was excavated to check its dimensions. The theory used was proposed on the basis of laboratory model tests by Hughes and Withers (1974). The purpose was to verify the theory on a field scale. A standard site investigation supplemented by Cambridge and Ménard pressuremeter tests provided basic soil parameters. The ultimate column load depends on the friction angle of the gravel used to form the column, the size of the column formed and the restraint of the clay on the uncemented gravel. To predict the load–settlement curve the essential radial stress–strain data for the clay were obtained from a Cambridge pressuremeter. The prediction is excellent if allowance is made for transfer of load from column to clay through side shear and correct column size. Accurate estimation of column diameter is the major factor influencing the calculation of ultimate load and settlement characteristics. The column improved substantially the bearing capacity of the natural soil.

La relation charge–tassement pour un essai à la plaque d'une colonne en pierre, isolée, dans l'argile molle, a été prévue avant l'essai in situ. Suite à l'essai, la colonne construite par la méthode de vibro-remplacement, a été excavée afin de vérifier ses dimensions La théorie utilisée est à la base d'essais sur modèle de laboratoire, d'après Hughes et Withers (1974). L'objectif a été de vérifier cette théorie à l'échelle du chantier. Une reconnaissance classique du terrain, complétée par des pressiomètres de Cambridge et Ménard, ont fourni les caractéristiques du sol. La charge limite de rupture d'une colonne, dépend de l'angle de frottement du gravier qui la forme, de sa taille et de la contrainte de l'argile sur le gravier non-cimenté. Afin de prévoir la courbe de charge–tassement, l'information essentielle de contrainte–déformation, a été obtenue à l'aide du pressiomètre de Cambridge. La prévision est excellente, si l'on tient compte du transfert de la charge, de la colonne à l'argile, grâce au cisaillement latéral et la taille exacte de la colonne. Une estimation précise du diamètre de la colonne, est le facteur principal pouvant influencer le calcul de la charge de rupture et du tassement. La colonne a considérablement amélioré la force portante du sol réel.

This Paper compares the field load bearing capacity of a stone column in soft clay at Canvey Island with the prediction of its behaviour made several weeks previously.

The stone column was constructed for plate loading to test a predictive theory on a site where Cementation Ground Engineering Limited was installing some 1000 similar columns by vibro-replacement for the foundations of two 31 m diameter oil tanks. Columns were formed by water jetting a vibroflot into the ground and compacting an imported gravel into the resulting hole as the vibrator was withdrawn. This standard replacement technique for strengthening soft clays has been described by Greenwood (1970, 1972).

* Senior Lecturer, Engineering Department, University of Auckland, New Zealand.
† Engineer, G. Maunsell & Partners, London.
‡ Technical Director, Cementation Specialist Holdings Limited, London.

As the location of the isolated test column was not initially known, the prediction had to be generally applicable to any position in the vicinity of the tanks. The prediction was made using information from a conventional site investigation supplemented with the results of tests with the new in situ pressuremeter developed at Cambridge (Wroth and Hughes 1973) and the Ménard pressuremeter.

The analysis of the behaviour of isolated stone columns proposed by Hughes and Withers (1974) provided the basis of the prediction. This analysis was developed from observations of laboratory models, and required proving on a field scale.

SITE CONDITIONS

Canvey Island is a low lying island on the north bank of the Thames estuary. Geologically it is very young; soft alluvial clays interleaved with sandy lenses overlie compact Thames gravels found about 30 m below ground level. Nowhere is the island more than 5 m above high tide level.

The logs of two deep borings at the centres of the tanks 100 m apart, indicate that the nature of the soil, on a macro scale, is uniform over the site (Fig. 1). It consists here of 9 m of soft grey clay on top of 11 m of silty sand which overlies a further 5 m of clay all above the Thames gravel.

PARAMETERS REQUIRED FOR PREDICTION OF COLUMN BEHAVIOUR

The important parameters for the prediction of column behaviour are

(a) The undrained shear strength of the soil.
(b) The in situ lateral stress in the soil.
(c) The radial pressure/deformation characteristics of the soil.
(d) The angle of internal friction of the column material.
(e) The initial diameter of the column.

A conventional site investigation using both laboratory and in situ tests was carried out to assess the properties of the soft clay. In addition five tests were made with the Cambridge pressuremeter to measure the in situ lateral stress and the radial pressure–deformation properties of the soil. Also two boreholes were tested with the Ménard pressuremeter.

The shear characteristics of the upper clay determined from triaxial tests, the pressuremeters, an in situ vane, and a Dutch cone are shown in Fig. 2. The general trend is clear: there is a top crust about 1 to 2 m thick, then soft clay to about 7 m depth, at which level the sand or silt is found.

The in situ radial stress as measured by the Cambridge pressuremeter is shown in Fig. 3. In the surface crust this is roughly equal to twice the cohesion. At a depth of about 6 m it approaches the value which might be expected for a normally consolidated soil. Near the ground surface the relatively high strength of the crust contributes most to lateral restraint on the column, but at depth the dominant component of lateral resistance is the weight of the overlying soil.

The calculation of column strength was based on the lowest value of measured passive restraint σ_{rL} which would be expected to occur over the critical length of column. The critical length is defined as the minimum length at which both bulging and end bearing failure occur simultaneously.

On the basis of stiffness measured by the Cambridge device (Fig. 4) the top 9 m of soil was divided into three zones: an upper layer from 0 to 3 m, a central zone from 3 to 5 m and from 5 to 9 m. The properties of these zones were used for the calculations of column behaviour.

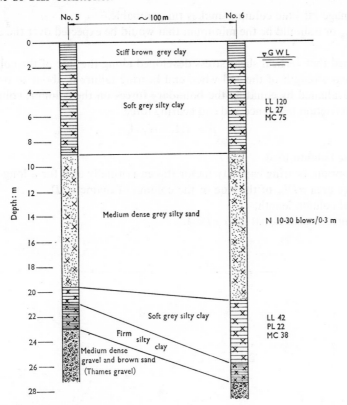

Fig. 1.   Logs of two bores on tank centres

The predictions were made for a cylindrical column which was assumed to have a diameter of 0·660 m.   This diameter was chosen on the basis of stone consumption on previous contract work for tank foundations in similar soil on a nearby site.

THEORY FOR ULTIMATE LOAD ON AN ISOLATED STONE COLUMN

The theory underlying these predictions is summarized in the following to outline the assumptions, from which the limitations of the analysis can be seen.

The results of quick pressuremeter tests elsewhere show that an acceptable approximation to the rigorous analytical expression for the total limiting radial stress σ_{rL} is

$$\sigma_{rL} \doteqdot 4c + \sigma'_{ro} + u_o \quad . \quad . \quad . \quad . \quad . \quad . \quad . \quad (1)$$

where σ'_{ro} is the initial radial effective stress and u_o the intial excess pore-pressure.   If a loaded column behaves similarly to a pressuremeter this expression may also be used for the lateral restraint of the soil on the uncemented column material.

If the stone in the column approaches shear failure with an angle of internal friction of ϕ' the limiting axial stress in the column is given by

$$\sigma_v = \sigma_{rL} \left(\frac{1 + \sin \phi'}{1 - \sin \phi'}\right) = \sigma_{rL} K_{ps}$$

or

$$\sigma_v = \left(\frac{1 + \sin \phi'}{1 - \sin \phi'}\right) (4c + \sigma_{r'_o}) \quad . \quad . \quad . \quad . \quad . \quad . \quad (2)$$

if $u_0 = 0$; drainage into the columns makes this probable.

The value of σ_{rL} or c should be the minimum that would be expected over the critical length of the column.

If it is assumed that vertical shear stress developed along the side of the column is equal to the average shear strength of the soil when end bearing failure is about to occur, the critical length can be evaluated by equating the boundary forces on the column; column load equals the sum of shaft friction resistance and end bearing force

$$p = \bar{c}A_s + N_c c A_c \qquad \dots \dots \dots \dots \quad (3)$$

where

p is the ultimate column load,

N_c is the appropriate bearing capacity factor (taken normally as 9 for a long column),

A_s is the surface area $\pi D L_c$ of the side of the column of diameter D,

L_c is the critical column length,

A_c is the column cross-sectional area $\pi D^2/4$,

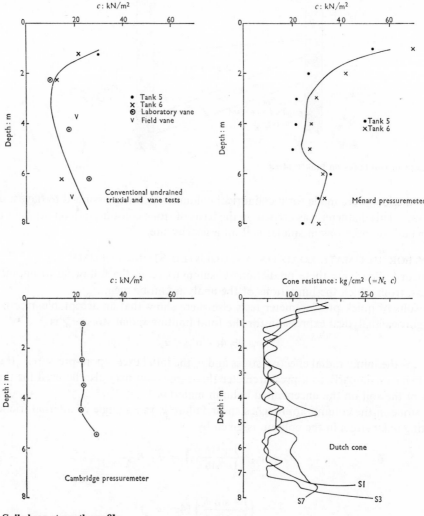

Fig. 2.   Soil shear strength profile

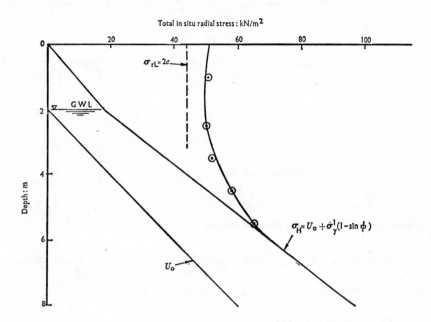

Fig. 3. Profile of in situ lateral stress from Cambridge pressuremeter

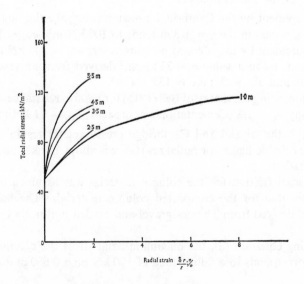

Fig. 4. Radial stress–strain curves from Cambridge pressuremeter

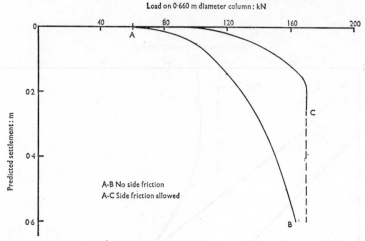

Fig. 5.   Predicted settlements of 0·660 m diameter column

$\bar{c}$ and c are respectively the average shaft cohesion and the cohesion at the bottom of the critical length.

Hence L_c can be determined by trial of values of soil properties.

PREDICTION OF ULTIMATE LOAD

The ultimate load is governed by bulging failure of the column in the upper zone.   It is the limiting radial restraint of the soil on bulging which determines the column load providing it cannot fail by lack of end bearing.

This restraint may be measured directly with a pressuremeter or calculated from cohesive strength and density of the soil as follows:

   (a) direct measurement by the Cambridge pressuremeter; the limiting pressure at which large expansion occurs in the upper 3 m tends to 100 kN/m²; $\sigma_{rL} = 100$ kN/m²,
   (b) direct measurement by the Ménard pressuremeter; $\sigma_{rL} = 200$ kN/m²,
   (c) by calculation, using a value $c = 22$ kN/m² derived from an assessment of site investigation data and $\sigma'_{ro} = 2c$; $\sigma_{rL} = 132$ kN/m²,
   (d) by calculation, using the theory of Bell (1915) for passive resistance with $c = 22$ kN/m² and bulk density $\gamma_b = 1·8$ (Cementation's approach); $\sigma_{rL} = 71$ kN/m².

Referring to Fig. 2, the triaxial and Cambridge pressuremeter results are similar.   Consequently the most probable figure for radial resistance is about 120 kN/m²: this value is used in subsequent calculations.

The angle of internal friction for the column material was assumed to be 38°.   This is believed to be reasonable for the compacted column material.   The backfill stone was a rounded river gravel derived from Thames gravel and graded uniformly between 20 mm and 40 mm.

On this basis, using equation (2), the maximum axial stress the column can withstand is 500 kN/m² which corresponds to a failure load of 170 kN on a 0·660 m diameter column.

CRITICAL COLUMN LENGTH

The critical column length is the shortest column which can carry the ultimate load regardless of settlement.   The basic assumption for this is that the soil/column interface develops the

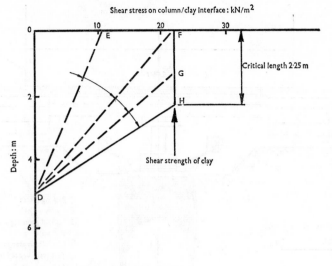

Fig. 6. Idealized build-up of shaft friction with column load

full cohesion of about 22kN/m² at failure.   Therefore by trial substitutions of values of $\bar{c}$ and c in equation (3) $L_o = 2.25$ m for a 0.660 m diameter column.

SETTLEMENT CHARACTERISTICS OF LOADED COLUMN

The stone columns as installed were 10 m long; they penetrated through the soft clay into much stiffer sand or silt.   For estimates of settlement of the test column its base was considered fixed at 7 m depth where the soil description changed.   Thus consideration of end bearing failure was excluded as its length was greater than L_o and the bearing soil was relatively strong.

The main assumption necessary for settlement calculation is that the column expands radially as settlement occurs retaining a constant volume.   To simplify the calculation the column is divided into layers and the total settlement is then related to the sum of the contributions from each layer
i.e.

$$\delta_v = \delta_1 + \delta_2 \cdots + \delta_n \quad . \quad . \quad . \quad . \quad . \quad . \quad . \quad . \quad . \quad . \quad (4)$$

where

$$\delta_n = 2H_n\delta_{rn}/r$$

where H_n is the thickness of the soil layer considered and δ_{rn}/r the radial strain for that layer (obtained from Fig. 4).

To relate radial strain to column load, a further assumption is required that the vertical to horizontal stress ratio in the column (K_{ps}) does not change significantly during settlement; and so the expected load–vertical deformation curve can be plotted for a given column diameter.

These assumptions are not strictly correct.   However, the inaccuracy they introduce is not significant in practice having regard to the accuracy which can be obtained in establishing soil and stone parameters.

Settlement was calculated in two ways.   The first method assumed that shear along the column/clay interface was negligible.   Hence the curve A–B in Fig. 5 could be calculated directly from equation (4) and Fig. 4.   The second approach made allowance for shear on the

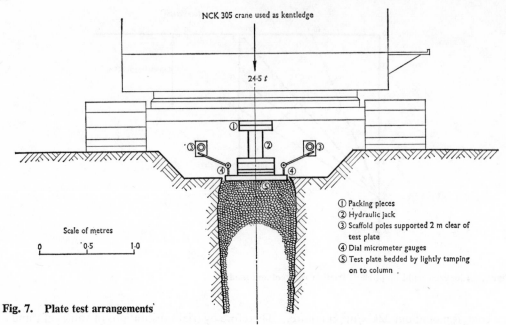

Fig. 7.   Plate test arrangements

① Packing pieces
② Hydraulic jack
③ Scaffold poles supported 2 m clear of test plate
④ Dial micrometer gauges
⑤ Test plate bedded by lightly tamping on to column

column/clay boundary by subtracting the total shear on the boundary from the total load above the horizon considered.   The assumptions were that

    (*a*) the column could not bulge at any depth unless the in situ lateral stress at that depth was exceeded and the full shear strength of the soil was mobilized on the column periphery,

    (*b*) as the column load increased, shear on the boundary between column and clay would build up until the maximum cohesion was developed at the top; with further increase in load the shear on the side of the column would progressively reach the maximum, dropping linearly to zero at 5 m depth (see line DGF in Fig. 6); at failure the full shear stress would be mobilized over the critical length (line DHF).

In contrast to the first method, the axial stress in the column decreases with depth as the axial load is transferred to the clay through shear.   The load–settlement graph derived by this method is AC in Fig. 5.   At failure the stress in the column at 2 m depth is 219 kN/m² which is barely enough to cause bulging; thus the deformation of the column will arise only above this level.   In both cases very little settlement is expected to occur if the load on the column is less than 60 kN.

RESULTS OF THE FIELD TEST

The column was tested by loading a concentric circular plate of 660 mm diameter—which proved to be marginally smaller than the top of the column.   Testing arrangements are detailed in Fig. 7.   The test took half an hour to complete and so it can be assumed that the soil deformed under undrained conditions.   Fig. 8 shows the variation of settlement with load. The observed failure load appears to be about 30% higher than the prediction in the previous section and 120% higher than predicted by Cementation.   The column also appears to be much stiffer than expected.   The method proposed by Hughes and Withers (1974) for calculating the ultimate load, although an improvement on current design, apparently underpredicts the allowable load by a surprisingly large amount.

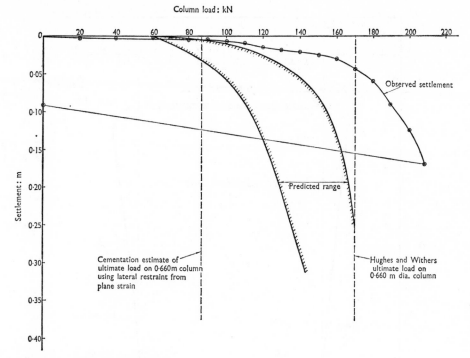

Fig. 8. Load test result with predictions

EXAMINATION OF TEST COLUMN

After the test, the column was subsequently excavated to determine its deformed shape. The results of those measurements are shown in Fig. 9. It is interesting to compare this deformed shape with that observed by Hughes and Withers (1974) in their model tests. The result of one of their tests is shown in Fig. 10. Clearly both the deformed shapes are geometrically very similar. Furthermore, the bulging, as predicted, was confined to the upper zone.

This suggests that the overconsolidated crust has either softened a little during installation of the columns or, more likely, as indicated by the field tests shown in Fig. 3, it does not provide the major resistance to bulging which might have been expected at first sight.

However, in practice the upper metre or two of column is often found to be shaped like a bucket, resting on a cylindrical stem through lower levels. This is due to the severe erosion of the top of the hole as each load of gravel is dumped around the vibroflot and falls into the bore. Further wear occurs as the machine is surged to promote flow of backfill and to assist compaction.

At Canvey the average column diameter calculated from the quantity of gravel used was 730 mm for over 1000 columns of 10 m depth. When the test plate was bedded down the test column diameter at that level averaged about 760 mm. While deformation during the test subsequently increased this, the various approximate measurements or calculations of diameter were mutually consistent.

It appears that the initial size of the real column was greater than the 660 mm assumed. As the failure load for a given failure stress varies with the square of the column diameter the estimation of the initial column size is of great importance.

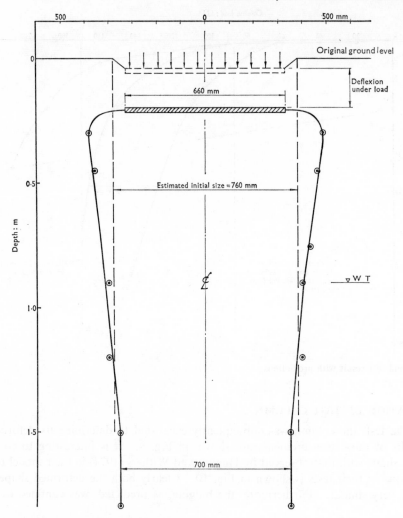

Fig. 9. Shape of column excavated after testing

If it is assumed that the column deformed at constant volume and that the upper section was initially a cylinder, then this would have an initial diameter of about 760 mm. This agrees fairly well with the measured surface diameter.

Using the average initial diameter of the upper 2 m, 730 mm, the estimates of the ultimate column load can be recalculated.

RECALCULATION OF FAILURE LOAD AND SETTLEMENT

When recalculating ultimate column capacity for 730 mm diameter the same value of stress of 500 kN/m² was taken based on $c = 22$ kN/m² as in the original calculation. It was assumed that the stress is uniform across any cross-section of the column; this must be very nearly true at only a short distance below the plate because of the close similarity of column and plate diameter. The revised estimates are plotted on Fig. 11. If the ultimate column load is calculated for each of the values of lateral restraint determined independently by the various

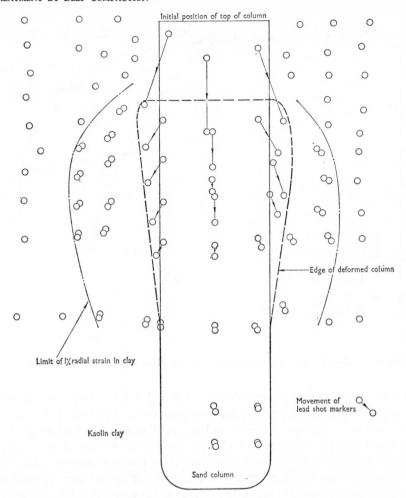

Fig. 10. Deformation of laboratory model column 38 mm diameter for vertical displacement of 25 mm (after Hughes and Withers, 1974)

site investigation techniques used, the results shown in Table 1 are obtained; they compare with the observed value tending to 220 kN.

Predicted settlement has also been recalculated for the revised diameter making the same allowances for side friction as before. The plot is shown on Fig. 11. The surprising agreement now obtained illustrates the importance of having an accurate estimate of column diameter for predictions.

REVIEW OF TEST IN RELATION TO COLUMN DESIGN

The results obtained confirm that the concept of the way in which stone columns improve bearing capacity of soft clays is correct. The remarkable similarity of shape of predicted and measured load–settlement curves, formerly in model studies and now in this field test, implies that the shear transfer envisaged alongside columns is valid. The value of pressuremeters for assessing soil restraint on columns is shown: the column load–settlement relationship cannot be established without the radial stress–deformation data from a Cambridge type pressuremeter.

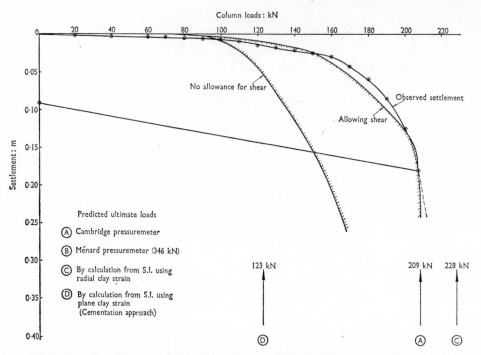

Fig. 11.   Predicted load–settlement characteristic for 730 mm diameter column compared with observed result

However, the importance of adopting correct soil and column properties for prediction of ultimate column load is also demonstrated.   An assessment of the major factors is pertinent.

The undrained shear strength of the soil appears in all methods requiring calculation of the lateral restraint to column bulging.   At Canvey, pressuremeters and in situ vanes suggested strengths almost double those assessed from laboratory tests on samples.   The higher value was used for calculations giving a lateral resistance similar to that directly measured by the pressuremeters and subsequently supported by the field column test.

In the absence of reliable in situ data this can be a major source of error, but likely always to be on the conservative side.

Radial restraint calculated from the usual triaxial test data in site investigation reports is also likely to be underestimated with respect to the effects of weight of overlying soil.   Direct measurement by pressuremeter in situ appears to be more reliable.

A major reason for underestimation of column load by Cementation's method is the use of Bell's theory for passive resistance.   This assumes plane strain and is very conservative for this application.   The difference is about 50%.

Table 1

Investigation technique	Ultimate column load, kN
Cambridge pressuremeter	209
Ménard pressuremeter	346
Calculation from site investigation	228
Cementation approach	123

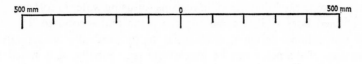

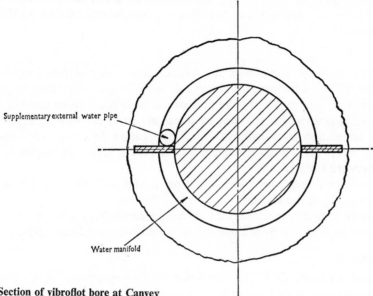

Supplementary external water pipe

Water manifold

Fig. 12. Section of vibroflot bore at Canvey

At Canvey the angle of internal friction ϕ' for the column material was assumed as 38°. The choice of uniformly graded gravel was to facilitate placing through the narrow annulus around the vibroflot to the bottom of the 10 m deep holes. If the assumed value of ϕ' is incorrect by up to 2° the effect on calculated column load is only about 6% variation. This is not a very important source of error.

The major error in the prediction of the test result was underestimation of column diameter. Although based on previous experience at the same site, some variation naturally occurs and the test column was slightly larger than assumed. Also, the calculated average diameter of all columns was larger than before.

The estimation of minimum column diameter is relatively easy if it is taken to be that of the vibroflot, but this would be unnecessarily conservative. Much depends on whether the column is constructed dry or, as at Canvey, with a water jet. In the former case the operation may be termed vibro-displacement as the vibrator forces its way into the soil. The resulting bore closely matches the dimension and shape of the machine. Vibro-replacement using water to transport soil from the bore produces an annular space round the machine. The vibroflot cuts a hole of diameter a little larger than that of its extremes across any fins or other appendages (Fig. 12).

At Canvey the vibroflots used were 410 mm diameter with two diametrically opposed fins giving an extreme diameter of 660 mm. Since the machines hang freely in the bore and when heavily taxed tend to rotate a little, a near circular bore is formed of diameter 50 mm or so larger than the fins. Water returning up the annulus does not have a powerful scouring effect because for a given flow volume very little increase in diameter is required for a considerable drop in velocity and cutting capacity of the water.

It should be remarked that the main reasons for using water in soft clays are first to maintain a stable unlined bore whilst backfilling the stone and second to form the largest possible column. For column design it is essential to be prudently conservative but even modest underestimates of diameter can be excessively so. Finally, it is noted that at Canvey the improvement of bearing due to the column compared with a plate at the same level on natural clay was about 2·5 to 4 times, depending on the value of c chosen from the data for the estimate of plate bearing capacity.

CONCLUSION

The explanation of the behaviour of a loaded stone column in clay proposed by Hughes and Withers (1974) has led to an improved predictive design method. However, the variations created by uncertainties in estimates of column diameter leave much to be desired.

Determination of lateral passive restraint by direct measurement with a pressuremeter is preferable to its calculation from standard site investigation data; this removes another potential source of error. If column settlement is to be predicted the radial stress–strain data from pressuremeters are essential.

The improved design theory now requires extension and field checking for widespread loading and for strip or pad foundations.

The work reconfirms model experience that at shallow depths very substantial increases of bearing capacity can be obtained from stone columns well compacted into clays.

ACKNOWLEDGEMENT

The Authors wish to thank Dr C. P. Wroth for his encouragement of this project, and Cementation Ground Engineering Limited for undertaking much of the field work and making results available.

REFERENCES

Bell, A. L. (1915). The lateral pressure and resistance of clay and the supporting power of clay foundations. *Proc. Instn Civ. Engrs* **199**, 233.
Greenwood, D. A. (1970). Mechanical improvement of soils below ground surface. *Proc. Ground Engineering Conf. Instn Civ. Engrs*, June, 11–22.
Greenwood, D. A. (1972). Vibroflotation—rationale for design and treatment. *Symposium on Methods of Treatment of Unstable Ground, Sheffield Polytechnic*, Sept.
Hughes, J. M. O. & Withers, N. J. (1974). Reinforcing of soft cohesive soils with stone columns. *Ground Engineering*, May, 42–49.
Wroth, C. P. & Hughes J. M. O. (1973). An instrument for the in situ testing of soft soils. *Proc. 8th Int. Conf. Soil Mech. Fdn Engng Moscow* **1**, 487–494.

Some applications of the vibro-replacement process

E. RATHGEB* and C. KUTZNER*

*The vibro-replacement process is a recognized foundation technique, which permits improvement
of in-place soils with an appreciable content of fines, with a view principally to their bearing
capacity and settlement properties, and to their shear strength.   As the name implies, the natural
soil is partially replaced by granular material infilled into holes made with the depth vibrator and
compacted. Two examples are shown.   In the first, the primary objective lay in the increase of
bearing capacity and the reduction of deformability of the foundation soil of a large thermal
power plant in a coastal region.   The second example refers to the increase of shear strength in a
soft layer on which an embankment for a motorway was to be placed.   In this case the use of a
computer program enables one to find the most economical number and arrangement of the stone
columns according to a given factor of safety in the stability analysis.*

*Two varieties of the process came to be applied, namely the wet and the dry processes.   Also
referred to is the case where that part of pile loads carried by skin friction is increased by this
method in order to reduce point loads.   Under certain, but nevertheless widely encountered, soil
conditions the vibro-replacement process is a technically sound and an economically interesting
alternative to piled foundations, proven over a period of nearly twenty years.*

*La méthode de vibro-remplacement est une technique de fondation reconnue, permettant une
amélioration de sols in situ ayant un pourcentage appréciable de fines et concernant principale-
ment leur force portante, leur propriété de tassement et leur résistance au cisaillement.   Une
partie du sol naturel est remplacée par des matériaux pulvérulents insérés dans les trous faits par
le pervibrateur qui les compacte.   Deux exemples sont cités.   Dans le premier cas, le but
principal a été d'augmenter la force portante et de réduire la déformation du sol de fondation d'un
centre thermal important, situé dans une région côtière.   Le second cas se rapporte à l'augmenta-
tion de résistance au cisaillement dans une couche molle, sur laquelle avait été projeté un remblai
pour une autoroute.   Dans ce cas, l'utilisation d'un ordinateur a permis de déterminer le nombre
et la disposition les plus économiques des colonnes en pierre, compte tenu d'un facteur de sécurité
donné dans l'analyse de stabilité.   Deux procédés ont été appliqués: la méthode mouillée et
la méthode sèche.   On note également le cas où la partie des charges de pieux supportée par le
frottement superficiel est augmentée par cette méthode, afin de réduire les charges ponctuelles.
Il a été prouvé sur une période de presque vingt ans, que dans certains types de sols, rencontrés
assez fréquemment, la méthode de vibro-remplacement présente une alternative techniquement
valable et économiquement intéressante pour des fondations sur pieux.*

The soil replacement process was developed in the late 1950s as a further application of the
deep compaction machine originally conceived for the densification of natural deposits or
made fills of loose sands and gravels.   The process has often been described in technical publications and can be considered as an accepted form of foundation technique (Greenwood,
1970; Keller, 1967; Webb and Hall, 1969).   In the following text some cases will be described
where this technique was used to solve difficult foundation problems very satisfactorily.

* Johann Keller GmbH, Frankfurt am Main, German Federal Republic.

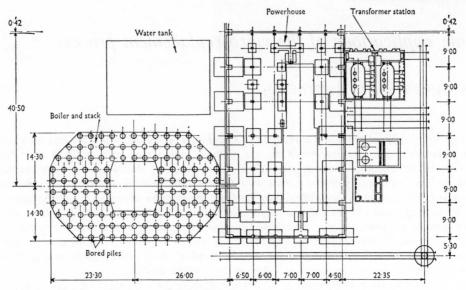

Fig. 1. Foundation plan of section of power station

The first case refers to the improvement of foundation soils to obtain increased bearing capacity and to reduce settlement.   The soil in coastal regions close to river estuaries is usually very heterogeneous and is generally not capable of supporting heavy structures without remedial treatment or the recourse to deep foundations.

Thermal power stations are often located in such regions and in the past 15 years several have been built on ground improved by the vibro-replacement process.   The structural characteristics of thermal power stations usually make shallow foundations preferable to deep foundations, for economic and other reasons, with the possible exception of very heavy boilers and stacks.   Such requirements can only be met if the soil can be simulated by a simple model with fairly well defined properties; a very heterogeneous soil has to be transformed artificially into statistically homogeneous layers to such depths where the pressures due to the imposed loads have dissipated sufficiently.   This can be effected very conveniently by the vibro-replacement process, as it is adaptable to individual situations and soil conditions.

Figure 1 shows a typical layout of a power station, composed of several such units built in various stages over a decade.   With the exception of the boilers and stacks all the structures have shallow foundations.   The soil (Fig. 2) consisted essentially of fine to medium sands with occasional gravels to a depth of about 16 to 18 m, irregularly interspersed with lenses of silt. The relative densities varied between loose and medium, according to the results of penetrometer tests.

This upper stratum is followed by soft sandy silts with inclusions of gravel and organic material, to depths between 30 and 32 m.   Then comes a layer, 6–7 m thick, of dense sand and gravel, underlain in its turn by medium clayey and silty soils to greater depths.   The groundwater table is approximately 3 m below ground level varying slightly with the adjacent tidal water.

The general level of the shallow foundations lies 2·5–3·0 m below ground level, only the foundations for the turbines and condensers being at greater depths.   The boilers and the stacks are supported by bored cast-in-place concrete piles which reach down more than 30 m into the layer of dense sands and gravels.

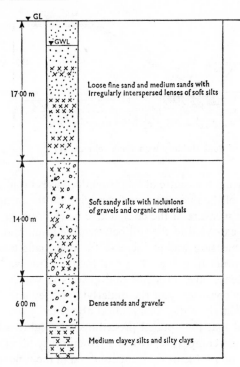

Fig. 2. Simplified soil conditions at site of power station

In order to be able to support the shallow foundations it was decided to treat by deep vibro-compaction all the areas corresponding to the powerhouses and transformer stations. In this way a deep block of stabilized soil was created within which all the foundation loads were contained. Due to the characteristics of the structures and of the soil, stabilization was extended 14·0 m below foundation levels in the powerhouses and to 5·0 m in the transformer stations and general areas. The shallow foundations were designed for allowable bearing pressures of 450 kN/m², with maximum edge pressures of 650 kN/m².

Compaction points were arranged in a triangular pattern with distances between points of 1·82 m, resulting in one compaction point for every 2·85 m² of treated area. In some critical areas additional intermediate points were placed. The so-called wet process was applied, where the vibrator was jetted into the ground with water and the silty material washed out by surging. After reaching the desired depth, crushed stone, 20 to 50 mm size, was filled into the hole against the throttled water flow. A stone column with widely varying sections was built up in lifts of about 0·8 m by retracting the vibrator partially and lowering it again to displace and compact the infilled stone.

Depending on the type of soil and on the volume of washed out silty material the take of infilled stones was between 0·3 and 0·7 m³/linear metre of column, with an average of close to 0·5 m³ for the whole job.

Control of quality was based primarily on the intensity of the electric current consumed, whereby a certain minimum had to be achieved before proceeding to fill in and compact the next lift. Penetrometer tests carried out before and after the treatment showed a marked increase in number of blows in those soils where the content of fines was small, while little or no increase was noted in soils with a higher content of fines. Such a result was to be expected

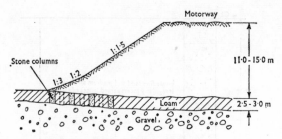

Fig. 3.   Example of increase of slope stability with stone columns (1)

as the radius of influence from the centre of vibration decreases rapidly with increasing content of fines.   This fact is one of the main reasons for relying on the measurement of the intensity of the electric current as a measure of the compactive work done.

The treatment was also extended to the soil between the foundation piles of the boilers and stacks.   The aim of this measure was to increase the pile loads taken up by friction in the upper part of the piles in order to reduce the point loads.   This was necessary in view of the limited thickness of the dense layer of sands and gravels in which the piles took their end-bearing and the rather low consistency of the underlying clayey and silty soils.

Measurements carried out since the construction about ten years ago, of the first stage of the power station, showed that the applied procedure was highly satisfactory.   Total settlement was kept within values compatible with the structural requirements and the differential settlements were such that no problems were encountered in the functioning of the installations of the power station.

Apart from the improvement of soils by vibro-replacement in view of settlement and bearing capacity of shallow foundations, the stone column process also has an important field of application when the improvement of shear strength of soft soils is the predominant problem. Such an improvement is due to the inclusion of highly compacted stone within the soft soils,

Table 1.   Assumptions and results of calculation of slope stability for embankment of Fig. 4. Soil values of ϕ and c are 0° and 30 kN/m² respectively; stone column values of ϕ and c are 45° and 30 kN/m² respectively

Line	Situation		Critical radius, m	Circle entre, No.	Factor of safety η
1	Pos. C		17·5	6	1·97
2	Pos. B		19·7	6	1·52
3	Pos. A		21·9 (25·9)	6 (10)	1·33 (1·33)
4	Pos. A′		21·9 (25·9)	6 (10)	1·37 (1·37)
5	Pos. A″		17·9 (21·9)	1 (5)	1·36 (1·36)
6	Pos. ′A		21·9 (25·9)	6 (10)	1·28 (1·28)
7	Pos. ″A		21·9 (25·9)	6 (10)	1·23 (1·23)
8	Pos. D		21·9 (25·9)	6 (10)	1·32 (1·32)
9	Pos. E		21·9 (25·9)	6 (10)	1·32 (1·32)
10	Pos. F		25·9 (27·9)	10 (6)	1·37 (1·38)
11	Untreated soil		21·9 (17·9)	6 (2)	1·03 (1·03)

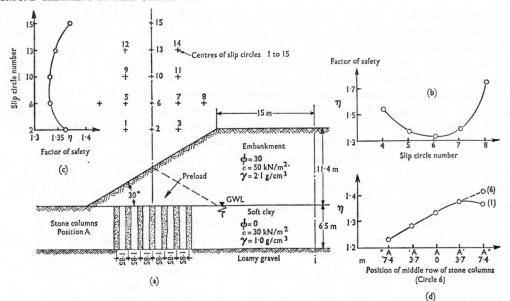

Fig. 4. **Embankment on soft layer and results of stability analyses for slip circles 1–15 with soil impr ovement**

forming a sort of reinforcement, and to the stone columns acting as drains through which the excess pore-water pressure in the surrounding natural soil can dissipate.

Knowledge regarding the shear strength of the stone columns themselves, as well as the composite material stone column plus surrounding soil is based on experience and on data from small and large scale tests. The increase of shear strength of soft soils treated by the stone column process leads to an interesting application which has been carried out for several years and has proved itself a technical and economic success.

Figure 3 shows an embankment that was built about ten years ago to a height of up to 15 m over a layer of soft loam, 2·5 to 3·0 m thick, underlain by dense gravel and sands (Keller, 1967; Schulze, 1962). Stability analyses had led to the conclusion that the shear strength of the loam would have to be increased if slopes of 1:3 to 1:1·5, adopted for economic and other reasons, were to be maintained. It was therefore decided to improve the shear strength of the loam in such a way that the calculated factor of safety of the slope of the embankment could be increased to 1·3, which was considered to be sufficient.

In such stability analyses it was then necessary to adopt an average value for the shear strength of the treated soil, in order to maintain within reasonable limits the calculation effort involved. Such average values were chosen very conservatively on the basis of past experience. It is furthermore well known that even a small value of the angle of shearing resistance has a remarkably favourable influence on the factor of safety when the inclination of the slope is less than about 40° (Terzaghi, 1943), which is usually the case. The assumption of an average shear strength could therefore not lead to a great error. Modern calculation methods have permitted analysis of such cases with more certainty and precision, as will be shown in the following example, the basis of which is shown in Fig. 4.

A few years ago an embankment for a motorway was to be placed on a layer of very soft clay with organic enclosures, underlain by loamy gravel and Keuper marl. The embankment was to have a maximum height of 11·4 m and the layer of clay was up to 6·5 m thick. In-place vane tests showed the clay to have an undrained shear strength of about 40 kN/m². Other geotechnical properties are given in Fig. 4.

Stability analysis for the case of untreated clay layers gave factors of safety of less than unity, whereupon a treatment with stone columns was envisaged. Tenders were called for a design of the improvement in order to ensure a factor of safety of at least 1·3. This particular case led to extensive theoretical considerations and to subsequent establishment of a computer program which allowed for variations in shear strength of the treated soil, and the influence of the individual stone columns or rows of columns. This program enabled the most favourable disposition of the stone columns to be determined. The program is based on the conventional analysis with circular failure lines considering the lateral forces on the individual slices. The analysis can also include seepage forces in the embankment and foundation soils.[1]

In the present case ten different arrangements of the stone columns within the improved soil were analysed with fifteen slip circles each, as shown in Fig. 4(a) and Table 1. The most economical solution consisted of seven rows of columns placed under the slope at 1·85 m centres transverse to, and 1·70 m centres parallel to the toe of the slope.

Firstly, the analysis with the slip circles 1 to 15 gave the critical circle centred on point 6. Then it was shown by the calculation of position B and C with different radii from point 6 that the critical circle was tangential to the boundary between the soft layer and the underlying dense material (positions A, B and C in Table 1). These two facts are known and theoretically demonstrated for the case of a slope of homogeneous cohesive soil on a firm base (Terzaghi, 1943), but the general validity of these relationships could not be initially presumed, as in the present case there is vertical as well as horizontal stratification. The position of the centre of the critical circle was finally established between the centres 6 and 10 (Fig. 4(c)).

The effect was investigated of a displacement of the seven rows of stone columns towards the toe of the slope and towards the crest of the embankment (positions "A, 'A, A, A' and A" in lines 3 to 7 of Table 1). The numerical result is shown in Fig. 4(d). In position A" the critical slip circle 1 did not cut through all the stone columns, so that not all of them contributed to the stability; however, for all other cases the slip circle 6 was determinant. For position A" the critical circle was displaced towards centre 1. For circle 6 a substantially higher factor of safety (1·41) resulted.

Finally, positions D, E and F were investigated, in which the distances between stone columns were increased to 2·4 and 3·6 m (positions D, E), or touching each other (position F). For positions D and E lower factors of safety were obtained than for position A; for F, the factor of safety was naturally higher but such a close spacing was unobtainable in practice.

In the project described the height of the embankment was variable; Fig. 4 shows the most unfavourable case. At smaller heights of the embankment the number of rows of stone columns was reduced, but the same geometrical arrangement was maintained. About 6200 stone columns with a total length of about 31 200 m were built up by the dry process. In this process, the vibrator is introduced into the soil without water jetting and to fill in with stones, in lifts, the vibrator must be fully retracted each time.

REFERENCES

Greenwood, D. A. (1970). Mechanical improvement of soils below ground surface. *Proc. Conf. Ground Engineering, ICE, London.*
Keller, J. (1967). Company brochure. Frankfurt/Main: Johann Keller.
Schulze, G. (1962). Ingenieurgeologische Untersuchungen im Untergrund einer Dammschüttung. *Zeitschrift der Deutschen Geologischen Gesellschaft* **114.**
Terzaghi, K. (1943). Theoretical soil mechanics. New York: Wiley & Sons.
Webb, D. L. & Hall, R. I. V. (1969). Effects of vibroflotation on clayey sands. *Jnl. Soil Mech. Fdn Div. Am. Soc. Civ. Engrs* **95,** SM6.

---

[1] The program was performed by Dr Frank Rausche,Cleveland—formerly member of Joh. Keller GmbH, Frankfurt.

Performance of an embankment supported by stone columns in soft ground

J. M. McKENNA,* W. A. EYRE† and D. R. WOLSTENHOLME‡

To investigate the effectiveness of stone columns in reducing the settlement of high embankments built on soft alluvium, stone columns were constructed under one end of the East Brent trial embankment using the vibro-flotation replacement technique. Here the alluvium was 27·5 m thick, the columns were 0·9 m in diameter and 11·3 m long, and they were constructed on a triangular grid at 2·4 m centres. The embankment was built to a height of 7·9 m. The foundations were instrumented, and a comparison of the piled and unpiled ground shows that the columns had no apparent effect on the performance of the embankment. Some of the construction problems are commented on, and the performance records and possible reasons for the ineffectiveness of the columns are presented in the Paper.

Afin d'enquêter sur l'efficacité de colonnes en pierre dont l'objet est de réduire le tassement de hauts remblais posés sur un dépôt alluvionnaire mou, de telles colonnes ont été construites au-dessous d'une extrémité du remblai d'essai de East Brent en utilisant la technique de remplacement par vibro-flottation. Dans ce cas particulier le dépôt alluvionnaire avait une épaisseur de 27·5 m; chaque colonne ayant 0·9 m de diamètre et 11·3 m de longueur. Elles furent disposées selon une maille triangulaire à 2·4 m des centres. On a construit le remblai à une hauteur de 7·9 m. Les fondations ont été pourvues d'appareils de mesure et on a constaté en comparant le terrain sur pieux et celui sans pieux, que les colonnes n'ont aucun effet prononcé sur la performance du remblai. Plusieurs problèmes de construction sont discutés dans le document ainsi que les détails de performance et les diverses raisons auxquelles on peut attribuer l'inefficacité des colonnes.

The design of the motorway across the Somerset Levels in Southwest England required embankments about 9 m high to be built on 27 m of soft alluvium. To study the performance of high embankments on soft ground, it was decided, before finalizing the design, to build a trial bank near East Brent (Map Reference ST 357528). Following discussions with the Cementation Company Ltd, stone columns were installed under one end of this embankment before fill placing started. The information relating to the performance of the stone columns is given in this Paper. Fig. 1 shows the trial bank in plan and section.

GEOLOGY

The Somerset Levels, according to Green and Welch (1965), consist of Pleistocene to recent estuarine alluvial sediments which were laid down in two major depositions. The first, from 6300 to 5000 BC, due to a relatively rapid eustatic rise in sea level, deposited initially sandy, then grey estuarine clay with minor peat beds up to about Datum Level. It was followed by a long period of peat accumulation over a wide area. This is the main peat bed and it is generally found at Ordnance Datum. Another marine incursion at about AD 250 deposited

* Consulting Geotechnical Engineer, Kingston Hill, Surrey.
† Chief Resident Engineer, Freeman Fox & Partners, Bristol.
‡ Senior Engineer, Freeman Fox & Partners, London.

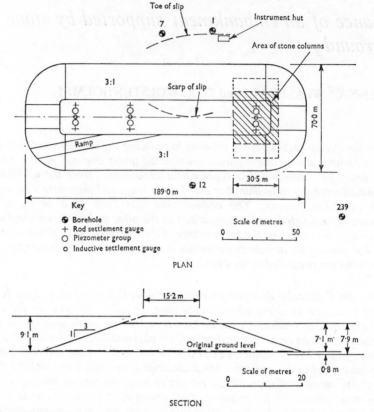

PLAN

SECTION

Fig. 1.   East Brent trial bank: plan and section

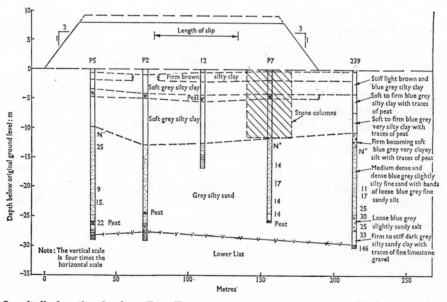

Fig. 2.   Longitudinal section showing soil profile

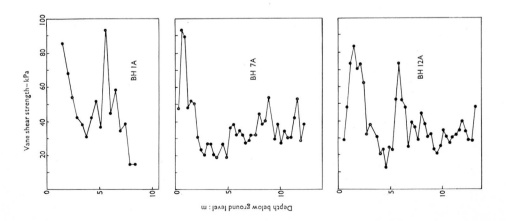

Fig. 3 (left).   Vane strength results

Fig. 4 (above).   General view of site during construction of stone columns

Fig. 5.   Forming holes

Fig. 6.   Backfilling holes

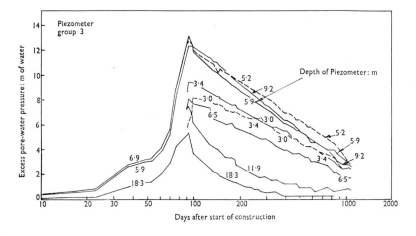

Fig. 7.   Pore-pressures in alluvium under centre of trial bank

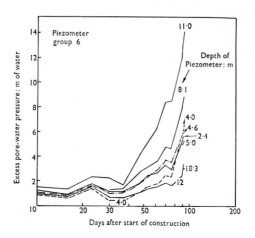

Fig. 8.   Pore-pressures in stone column zone of alluvium

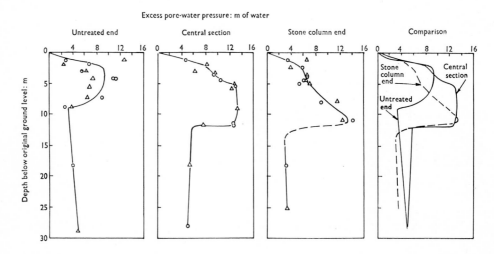

Fig. 9.   Pore-pressures on day prior to slip

a blue-grey estuarine clay about 4·3 m thick over the peat with a flat upper surface 5·5 to 6·1 m above OD, the top of which has since been desiccated.

Hawkins (1971) presented an alternative hypothesis based on pollen and carbon dating and archaeological finds. This evidence indicates that there has been no recent major marine incursion and that since Roman times the land has accreted by only about 0·3 to 0·6 m.

At the site, about 12·5 m of soft clay with peat layers overlay 15 m of sand and silt. Fig. 2 shows a longitudinal section and the soil profile, and Fig. 3 gives vane strength results. The average vane strength of the soft clay is about 26 kPa, and the eight SPT results in the under-lying sand and silt vary between 9 and 25.

CONSTRUCTION OF THE STONE COLUMNS

After discussions with the Cementation Company Ltd, stone columns 0·9 m in diameter and 11·3 m long were constructed at an average spacing of 2·45 m in a triangular grid. A 30·5 m length of the foundations at one end of the trial bank was treated in this way. The 0·9 m holes were formed with two vibroflots, using a water jetting technique. Each machine required five litres of water per second during operation, and settling ponds were required to allow the slurry to settle out before the water could be returned to the drainage ditches (rhynes). Fig. 4 shows a general view of the site during construction of the stone columns, Fig. 5 a close-up view of the holes being formed and Fig. 6 the backfill being placed.

The intention was to treat the area above which the embankment would be more than 3 m high. Unfortunately, the rate of construction was slower than had been anticipated, and there was only time to treat about two-thirds of the area originally intended.

The holes were backfilled with a crushed limestone of nominally 38 mm single size.

INSTRUMENTATION

Three groups of instruments were installed in the alluvium, one in the centre of the stone piled zone and two in the untreated foundations (Fig. 1). Each group consisted of 16 piezo-meters, one inductive settlement gauge and three rod settlement gauges.

CONSTRUCTION OF THE EMBANKMENT

Construction of the trial bank started at the end of October 1967 and continued for three months. It was 189 m long, 70 m wide with side slopes of 3:1, and was built of compacted quarry waste with not more than 15% passing a No. 200 BS sieve, except for the bottom 460 mm where the fines were limited to less than 10% in order to form a drainage layer. After com-paction by at least 8 passes of a vibrating roller, the fill had a density of about 22·8 kN/m³. The intention was, if possible, to construct the bank to a height of 9·1 m with a final crest width of 15·2 m. However, with 7·9 m of fill in place, a 60 m length in the centre of the embankment slipped 92 days after the start of construction, and filling was stopped (McKenna, 1968).

PIEZOMETER RECORDS

The build-up in pore-pressures during construction and the subsequent dissipation under the centre of the bank are shown in Fig. 7. Fig. 8 shows only the build-up in pore-pressures in the stone piled area, as the leads from these piezometers were severed by the slip and there-fore there are no dissipation records. Fig. 9 shows the pore-pressures measured on the day before the slip for the three instrumented sections. It is apparent that the stone columns had little effect on the amount of pore-pressure build-up during the construction period. The highest construction excess pore-water pressure, 14·2 m, was in fact measured at a depth of 11 m in the stone column zone, but this was possibly due to load transfer.

SETTLEMENT RECORDS

Rod settlement gauges

The settlements for the three rod settlement gauges on the centre line of the bank are plotted against log time in Fig. 10. The settlement of the stone column area, in both amount and rate of settlement, was identical to that of the untreated central area.

Inductive settlement gauges

The inductive settlement gauge readings for the three sections on day 90 (two days before the slip) are given in Fig. 11. After the slip, the central gauge became inoperative, the other two gauges started giving trouble, and readings were taken with increasing difficulty up to day 188. The last readings are shown in Fig. 12. These records show that the settlement in the upper 12·5 m of alluvium was a uniform 440 mm on day 90 and 650 mm on day 188. The difference in the total settlement between 1380 mm in the stone column area and 830 mm in the other end was due to settlement occurring in the 15 m of alluvium below 12·5 m. It is interesting to note that at this time the settlement of 730 mm in the sands and silts below the stone columns was greater than the 650 mm in the clayey alluvium. It is apparent therefore that the stone columns did not reduce the amount of settlement or increase the rate of pore-pressure dissipation.

CONCLUSIONS

The instrumentation records showed that the stone piles had no effect on the amount or rate of settlement of the East Brent trial bank built on 27·5 m of alluvium, and as a result stone columns were not used under the motorway embankments.

It is postulated that these columns were ineffective for two reasons.

The grading of the 38 mm single size crushed limestone was too coarse to act as a filter, and as a result, the voids in the gravel backfill probably became filled with clay slurry which prevented them from acting as drains. In addition, the method of construction would probably have remoulded the adjacent soft clays and damaged the natural drainage paths, so nullifying any potential drainage provided by the stone columns.

Possible reasons why the stone columns did not reduce the settlement in the upper 12·5 m of clayey alluvium are not so obvious. One explanation might be that the backfill was so coarse that when the embankment load came on to the columns the crushed stone forming each column was not restrained by the surrounding soft clay, and as the columns expanded, the soft clay squeezed into the voids.

ACKNOWLEDGEMENTS

The Paper is presented by permission of Mr P. G. Lyth, Director of the South West Road Construction Unit, and of Mr W. T. F. Austin, the partner of Freeman Fox and Partners responsible for the Clevedon Hills and Mendip Hills Sections of the M5 motorway.

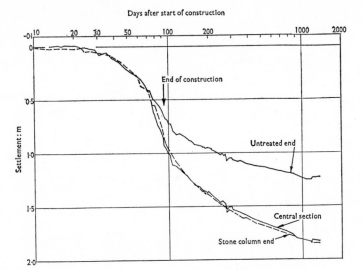

Fig. 10. Settlement of the three centre line rod settlement gauges against log time

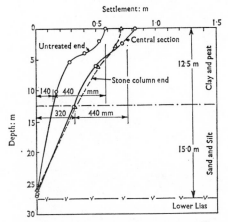

Fig. 11. Inductive settlement gauge readings on day 90, two days before the slip

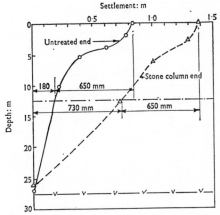

Fig. 12. Inductive settlement gauge readings on day 188

Many engineers were involved on site in taking all the readings during the construction of the trial bank and in the following three years, and the Authors gratefully acknowledge their enthusiasm and attention to detail. Particular mention should be made of Mr D. A. Cook who took the instrument readings during the construction of the trial bank.

The following Contractors were involved: the Cementation Company Ltd (stone columns), Soil Instruments Ltd (instrumentation), A. E. Farr Ltd (trial bank), Soil Mechanics Ltd (site investigation).

REFERENCES

Green, G. W. & Welch, F. B. A. (1965). *Geology of the country around Wells and Cheddar*, 121. HMSO.
Hawkins, A. B. (1971). Sea level changes around south-west England. *Colston Papers* **23**, 67–88. Butterworths.
McKenna, J. M. (1968). *Site investigation of a slip in the East Brent (Somerset) trial embankment.* Soil Mechanics Limited Report No. 5101/1. Unpublished.

Field testing to evaluate stone column performance in a seismic area

K. ENGELHARDT* and H. C. GOLDING*

For construction of a 16 mgd sewage treatment plant on predominantly deep, soft, cohesive soils in an area of highest seismic susceptibility, soil improvement with stone columns was considered as one foundation solution. While the deformation behaviour of subsoil improved with stone columns, under static vertical load, has been documented, little if any information is available on the performance of such improved soils under seismic loading conditions. For this reason, large scale field tests were performed to demonstrate that (a) in the process of stone column installation, sand lenses in the predominantly cohesive subsoil are sufficiently densified with respect to liquefaction potential; (b) the combined mass of stone columns and native, intervening soil develops sufficient shear strength to resist safely horizontal forces resulting from a ground acceleration of 0·25 g; (c) the stone column pattern which satisfied the shear and density requirements also provides an adequate load–settlement relationship.

Pour construire une usine de traitement de résidus (ayant une capacité de 16 millions de gallons par jour) sur des sols cohérents mous sur une grande profondeur dans une zone à grande séismicité on a envisagé d'améliorer le sol avec des colonnes de pierre. Alors que l'on connaît bien le comportement sous une charge statique et verticale d'un sol amélioré par des colonnes de pierre, il y a peu d'information concernant le comportement de tels sols sous les actions sismiques. Pour cette raison, on a réalisé des essais de chantier, importants, afin de démontrer que (a) par la méthode d'installation de colonnes de pierre, les lentilles de sable dans les sous-sols essentiellement cohésifs, sont suffisamment densifiées vis à vis de la possibilité de liquéfaction; (b) les colonnes de pierre et le sol originel intervenant ensemble, peuvent développer suffisamment de résistance de cisaillement pour résister en sécurité aux forces horizontales résultant d'une accélération terrestre de 0·25 g; (c) le modèle de colonne en pierre satisfaisant aux exigences de densité et de cisaillement présente aussi la relation charge–tassement correcte.

Improvement of soft cohesive soils for construction purposes by means of vibro-replacement or stone columns using a vibratory probe has been well established for the past 15 years. Many successful applications have proved that the method is a valuable addition to the field of special foundation systems. Although a rigorous solution for evaluation and prediction of the behaviour of soils improved by the vibro-replacement method has not yet been developed; case histories, extensive studies, and experience permit a conservative semi-empirical design approach (Greenwood, 1970; Thorburn and MacVicar, 1968; Hughes and Withers, 1974).

Generally, stone columns have been used in the support of basic foundation types, such as small isolated footings, strip footings and large raft foundations.

By partial replacement of the soft in situ soils with compacted granular material, the use of stone columns results in reduced settlements. The higher strength characteristics of these

---

* Vibroflotation Foundation Company, Pittsburgh, Pennsylvania, USA.

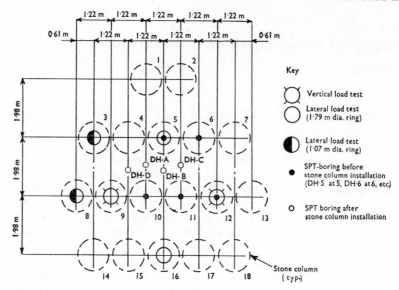

Fig. 1. Test field plan

compacted granular columns result in increased shear resistance for bearing capacity evaluation. Watt *et al.* (1967) describe stone columns as 'shear pins which increase the average
shearing resistance along any possible slip plane thus preventing failure'. The concept of
increased shear resistance suggested the use of stone columns for improvement of soft zones
under embankments where stability analyses indicated insufficient strength of the in situ
soils.

For evaluation of horizontal resistance it has generally been assumed that the stone column
develops an angle of internal friction of approximately 35° (Greenwood, 1970).

Given the specific geometry and the soil parameters of the native soil, and using an assumed shear strength of the stone columns, the total shearing resistance of the composite
mass of vertical, compacted granular cylinders with intervening native soil may be computed.
Relatively recent refinement of the calculation by computerization nevertheless is still dependent on the assumed angle of internal friction for the compacted gravel columns.

For the construction of a new 16 mgd sewage treatment plant on soft to medium stiff, fine-
grained estuarine deposits, soil improvement with stone columns was considered as one
foundation solution. For normal foundation analyses, the total shearing resistance would
have been of interest only in the calculation of ultimate bearing capacity. However, since
the plant was located in an area of highest seismic susceptibility, the measure of safety against
base shear failure (sliding) became one of paramount importance. The design criteria for
horizontal ground acceleration at foundation level had been established at 0·25 gravity.
Furthermore, the existence of sand lenses in the generally fine-grained estuarine sediments
raised the question of possible liquefaction of these sands during strong shaking.

In order to verify that the combined mass of stone columns and native intervening soil
developed the requisite shear strength and that the sand pockets in the predominantly fine-
grained soils were sufficiently densified to minimize liquefaction, a series of large scale field
tests was performed. The results of these tests are given in this Paper.

TEST PROGRAMME

As a location for the tests, an area at the future construction site was selected where the estuarine deposits extended to approximately 10·7 m below grade.

For a test field, 18 stone columns were installed in a 1·22 m × 1·98 m triangular grid pattern as shown in Fig. 1. Each stone column extended through the estuarine deposit and terminated in the underlying marine sands. A standard wet stone column method of installation was used (Engelhardt *et al.*, 1974). The backfill material consisted of natural and broken stones ranging in size between 19 mm and 76 mm. Based on measured backfill consumption, the average diameter of the stone columns was approximately 1·1 m.

The test programme consisted of the following phases.

(*a*) Standard penetration test borings before and after installation of stone columns.
(*b*) Two horizontal shear tests on stone columns alone and two similar tests on the combined system of stone column and contributory natural surrounding soil.
(*c*) Two vertical load tests.

The English system of measurement was used in the test programme but all units have been converted to the metric system for purposes of this Paper; this conversion accounts for certain awkward dimensions.

SPT AND LIQUEFACTION POTENTIAL

The layout of the test borings is shown in Fig. 1. The stratigraphy, as revealed by the test borings, and the SPT results are presented in Figs 2(a) and 2(b). Although the maximum distance between test borings was only 1·15 m, the difficulty of correlation between borings indicated the very erratic lenticular nature of the estuarine deposits. Underlying these estuarine sediments were medium dense to dense marine sands. The transition between the estuarine and older marine deposits was poorly defined.

Liquefaction is a phenomenon in which a cohesionless soil loses strength during an earthquake and acquires a mobility to permit movements ranging from several metres to several thousand metres (Seed and Idriss, 1971). It generally occurs in clean, poorly graded, loose sands with a mean grain size in the range of 0·1 to 0·2 mm. In addition to the soil type, other important factors are initial confining pressure, intensity and duration of ground shaking and relative density or void ratio of the sands.

Grain size analyses of the most sandy samples from the test site indicate that most of the sandy soils contained more than 20% passing the No. 200 sieve (0·074 mm) (Fig. 3). These sands are, by definition, not clean sands and are therefore less susceptible to liquefaction. Only one sample (borehole DH-D) showed relatively clean sands (18% passing sieve No. 200) at a depth of 7·9 to 8·8 m.

As shown in Fig. 2(a), the standard penetration test after installation of stone columns shows that this specific sand lense was significantly densified. Correlation with the SPT *N* values indicates a relative density greater than 92% which is in a range where liquefaction is very unlikely to occur (Seed and Idriss, 1971).

Lee and Albaisa (1974) have reported that clean sands were densified, by the vibroflotation method, beyond the critical range of liquefaction. The nature of the soil in the test area did not permit demonstration of increased densification of a relatively clean sand in more than one instance. However, using the same pattern and vibratory equipment as in the vibroflotation method, it can be reasonably inferred that clean sands, although interbedded with fine-grained soils, will also be densified to the same extent during the installation of stone columns.

SHEAR TESTS

The shear tests were designed to demonstrate the horizontal shear strength developed by a stone column improved subsoil. Direct quick shear tests were judged to be an acceptable simplification of the dynamic loading condition which an earthquake produces and where the soil is subjected to a series of reversing cyclic shear stresses and strains. Before testing it had been concluded that the plane of potential shear failure would occur high within the soft,

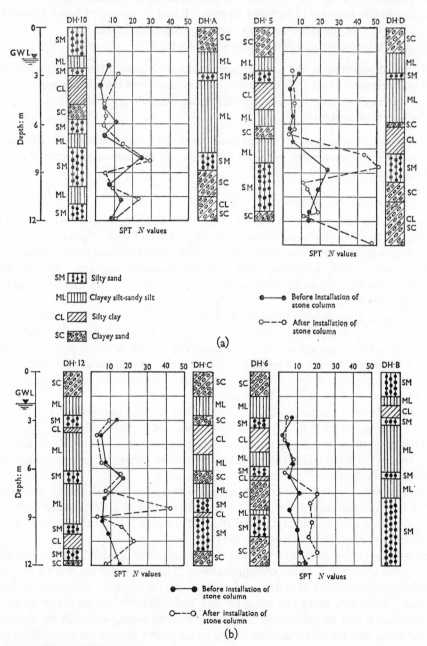

Fig. 2. Soil profiles and SPT *N* values

natural, cohesive soil underlying the existing fill. For this reason the test area was stripped of the fill, after stone column installation, in order to permit testing on material which had not been directly disturbed by working equipment during the stone column installation.

Two shear tests were performed with a 1·07 m diameter steel ring (area = 0·89 m²) confining only the stone column, and two tests were performed with a 1·79 m diameter steel ring (area = 2·5 m²) confining the area of one stone column and the contributory surrounding area of natural soil in accordance with the stone column pattern.

The two different sizes of shear area were selected to demonstrate, independently, the shear strength of the stone column alone and that of the combined mass of stone columns and natural soil. The shear resistance of the stone column alone is of importance since the most critical period for a shear failure through the foundation soil is immediately after

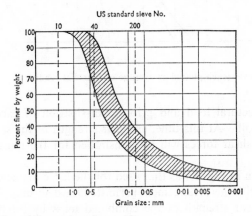

Fig. 3. Grain size distribution of the most sandy samples

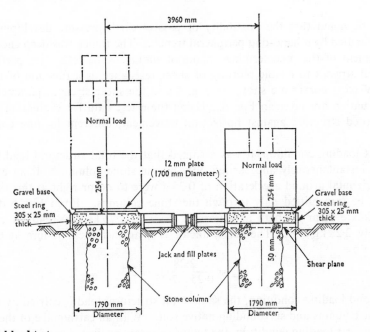

Fig. 4. Lateral load test

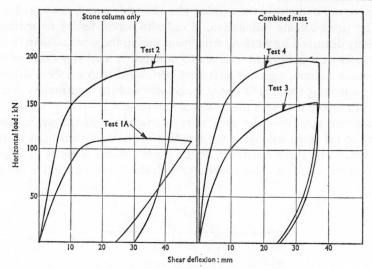

Fig. 5.   Load–deflexion curves (each plotted as average of four dial readings)

application of the structural load and prior to dissipation of excess pore-pressure in the soft, cohesive, native soil.   At this time, as shear strength of the saturated soil is limited to its cohesion only, the shear forces will be transmitted for the most part through the stone columns.

The test set-up shown in Fig. 4 was that used for the large diameter ring.   Performance of the test in accordance with this arrangement permitted the determination of the shear deformation behaviour and shear strength of the combined mass by forcing rupture along a predetermined horizontal plane.   The set-up was modified for testing of the stone column alone.

It should be noted that the possibility of passive earth pressure developing against the ring was eliminated by a hand-dug peripheral trench.   The trench was deep enough to permit free development of the predetermined plane of sliding.   The tests were performed at different vertical stresses to enable plotting of shear resistance as a function of normal stress. The load–deflexion curves are shown in Fig. 5 and the shear strength parameters using the Coulomb equation are given in Fig. 6.   Using these test data for evaluation of the safety of the proposed structure against horizontal movement, different loading conditions were studied.

In the first loading condition, it was assumed that the total structural load of $1\cdot0$ kg/cm² was applied instantaneously and carried only by the stone columns.   If an earthquake inducing a horizontal ground acceleration of $0\cdot25$ g were to occur right at this time, the shear force would be transmitted only through the stone columns.   From Fig. 6 it can be seen that the ultimate shear strength at a normal stress of $2\cdot75$ kg/cm² (240 kN on $0\cdot89$ m²) is approximately $2\cdot28$ kg/cm².   The safety factor (F_s) against shear failure at $0\cdot25$ g seismic force is given by

$$F_s = \frac{2\cdot28}{0\cdot25 \times 2\cdot75} = 3\cdot3$$

In the second loading condition, the same total structural load is carried by the combined mass of stone columns and surrounding native soil.   With an earthquake of the same magnitude as in the first loading condition, the factor of safety against shear failure is $3\cdot4$.

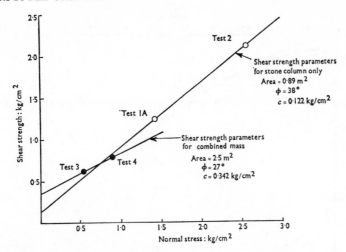

Fig. 6. Shear strength parameters (ϕ = angle of internal friction, c = cohesion)

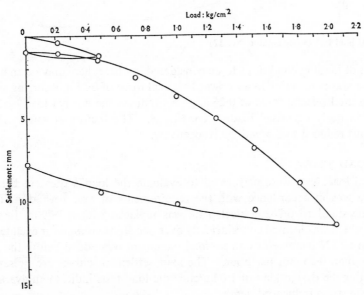

Fig. 7. Vertical load test (stone column No. 9)

The safety factors developed on the basis of the test results are more favourable than those calculated, before the tests, on the basis of laboratory test results and assumptions. As an example, for the loading condition where the combined mass carries the total structural load, the factor of safety initially calculated was 2·9. This factor was developed on the assumption that for a given depth, the settlement under a rigid foundation is the same for the stone column and the surrounding soil. The load distribution is then simply a function of the area and modulus of compressibility of the respective materials. For the stone column pattern of 1·22 m × 1·98 m, approximately 83% of the total load is carried by the stone columns and 17% by the native soil. The factor of safety was calculated using these load distribution factors, average laboratory shear parameters (angle of internal friction of 18° and a cohesion of 0·125 kg/cm²) and assuming an angle of internal friction of 35° for the stone column.

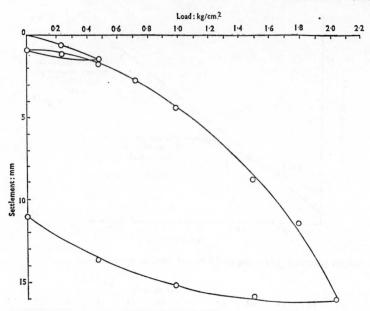

Fig. 8.   Vertical load test (stone column No. 12)

Another set of loading conditions to be considered is where there may be an upward propagation of shear waves resulting in an upward vertical force of 0·25 g occurring simultaneously with the horizontal seismic force of 0·25 g.   This reduces the normal load by 25% for each of the two previously discussed loading conditions.   The factors of safety which were 3·3 and 3·4 are thus reduced to 2·5 and 2·9 respectively.

VERTICAL LOAD TESTS

Two vertical load tests were performed to evaluate the load–settlement behaviour.   The test procedure was in accordance with the requirements of test method ASTM D1194-66 except that the standard 0·762 m steel plate was replaced with a 1·79 m diameter circular concrete slab which was located concentrically over one stone column for each test.   The load was applied in 62 kN increments and no load increment was added before the rate of settlement was less than 0·25 mm per hour.   The load–settlement curves are presented in Fig. 7 and Fig. 8.   For the design load of 1·0 kg/cm², the load tests indicate an average settlement of 4 mm for the stress influenced depth.

CONCLUSIONS

In a series of large scale field tests on stone columns installed with vibroflotation equipment, it has been demonstrated that

   (a) relatively clean sand lenses or layers interbedded with soft, cohesive soils are densified above the critical range of liquefaction,

   (b) the process of compacting gravel in a stone column results in a high angle of internal friction of 38°,

   (c) the shear strength parameters of the combined mass of stone columns and native, intervening soils are significantly higher than the parameters of the in situ soils existing prior to the stone column installation and that the resultant shear strength can safely resist horizontal forces induced by a ground acceleration of 0·25 g.

ACKNOWLEDGEMENTS

The Authors wish to thank Mr Ivar Staal, Geotechnical Consultants, Inc. and Professor J. K. Mitchell for their help in formulation and evaluation of the testing programme. The Authors also wish to acknowledge, with appreciation, the active participation of Mr A. A. Bayuk, Vibroflotation Foundation Company, in preparation of this Paper.

REFERENCES

Engelhardt, K., Flynn, W. A. & Bayuk, A. A. (1974). Vibro replacement, a method to strengthen cohesive soils in situ. *Am. Soc. Civ. Engrs. National Structural Engineering Meeting*, meeting preprint 2281.

Greenwood, D. A. (1970). Mechanical improvement of soils below ground surface. *Ground Engineering*, 11–22.

Hughes, J. M. O. & Withers, N. J. (1974). Reinforcing of soft cohesive soils with stone columns. *Ground Engineering*, 42–49.

Lee, K. L. & Albaisa, A. (1974). Earthquake induced settlements in saturated sands. *Jnl Geotechnical Engineering Division, Am. Soc. Civ. Engrs.*, 387–406.

Seed, H. B. & Idriss, I. M. (1971). A simplified procedure for evaluating soil liquefaction potential. *Jnl Soil Mech. Fdn Div. Am. Soc. Civ. Engrs* **97**, 1249–1274.

Thorburn, S. & MacVicar, R. S. L. (1968). Soil stabilization employing surface and depth vibrators. *The Structural Engineer* **46**, No. 10, 309–316.

Watt, A. J., de Boer, B. B. & Greenwood, D. A. (1967). Loading tests on structures founded on soft cohesive soils, strengthened by compacted granular columns. *Proc. 3rd Asian Reg. Conf. Soil Mech., Haifa* **1**, 248–251.

The role of ground improvement in foundation engineering

J. M. WEST, BSc, PhD*

Compaction techniques are increasingly used to solve a wide range of foundation problems and their scope now extends beyond the treatment of granular materials to include many silty or clayey soils. This development introduces the possibility of strengthening the ground on a much wider range of projects. It is argued that the idea of ground improvement has an important effect on traditional foundation selection procedures and can frequently lead to more economical foundation and substructure systems. The Paper indicates types of projects most likely to profit from this approach and quotes several examples where vibro-compaction and dynamic consolidation techniques were used to stabilize both granular and cohesive soils.

L'utilisation des techniques de compactage s'accroît afin de résoudre une vaste étendue de problèmes de fondation, et sa portée s'étend maintenant au delà du traitement de matériaux pulvérulents jusqu'à de nombreux sols limoneux ou argileux. Ce développement introduit la possibilité de renforcer le sol sur une gamme de projets beaucoup plus importante. Il est soutenu que l'idée d'améliorer le sol a un effet considérable sur les procédés traditionnels du choix des fondations et peut souvent conduire à des systèmes de fondation et structure plus économiques. Le document indique les genres de projets qui profiteront probablement le plus de cette possibilité, et donne plusieurs exemples où les techniques de vibro-compactage et de consolidation dynamique ont été utilisées afin de stabiliser les sols tant pulvérulents que cohérents.

In engineering practice the principal application of soil mechanics occurs in the selection of foundation type and construction method which, although a critical aspect of any project, is an area where cost effectiveness is at a particular premium. Increasing weight of structures and land utilization considerations constantly aggravate the problems inherent in foundation design and in response to these pressures many techniques have been developed to overcome, or at least minimize, the effects of adverse soil conditions.

The past decade has witnessed a substantial growth in the use of compaction techniques such as vibro-replacement, vibro-flotation and, latterly, dynamic consolidation; the point has now been reached where, far from offering solutions to a narrow range of unusual problems, these techniques find routine application on projects of widely varying character.

GROUND IMPROVEMENT

In the context of this Paper the term ground improvement embraces those special techniques, notably compaction techniques, capable of increasing the bearing capacity and decreasing the compressibility of natural soils or fills in situ. The practical value of this facility and its effect on foundation selection procedures can most easily be seen by considering the methods normally used in a routine problem.

* GKN Foundations Limited.

Table 1. Merit classification for ground improvement techniques relative to alternative methods

Type of structure	Structure	Permissible settlement	Loading density/ typical bearing pressure required, kN/m²	Probability of advantageous use of ground improvement techniques		
				Loose granular soils	Inert fills	Weak alluvia
Office/domestic	High rise: above 6 storeys	Low	High/300+	High	Low	Unlikely
Frame or load-bearing construction	Medium rise: 3–6 storeys	Low	Medium/200	High	Good	Low
	Low rise: 1–3 storeys	Low	Low/100–200	High	High	Good
Industrial	Large span portal and cranes, heavy machines, silos, chemical process plants	Low (differential settlement critical)	Variable with high local concentrations up to 400	High	Low	Unlikely
	Framed warehouses and factories	Fair	Low/100–200	High	High	Good
	Covered storage, storage rack systems, production areas	Low to medium	Low/up to 200	High	High	Good
Others	Road embankments	High	High/up to 200	Treatment unlikely to be required	Treatment unlikely to be required	Good
	Open storage areas	Medium to high	High/up to 250	Treatment unlikely to be required	High	High
	Storage tanks	Medium to high (dependent on dimensions, type and contents)	High/up to 250	Treatment unlikely to be required	High	High
	Effluent treatment installations	Medium (differential settlement important)	Low/up to 150	Treatment unlikely to be required	High	High

Note: Bearing pressure quoted is an estimate of the value required for efficient shallow foundation design.

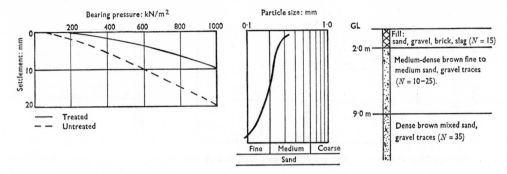

Fig. 1. High rise block; soil conditions and load test results

From the outset two key factors dominate the analysis; first, the structure, whether it be an embankment or a conventional building; and second, the site chosen for development. The structure implies certain loads and settlement tolerances while each site has its own geotechnical profile and the object in foundation design, as in all other aspects of the project, is to reconcile these at minimum overall cost. Most engineers assume that the optimum solution occurs when the applied loads can be supported at or near (say within 2 m) of the ground surface and consequently this is the first possibility to be examined. If successful, it is unlikely that alternatives will be examined unless soil support and groundwater conditions render even shallow excavation difficult or where the foundation dimensions become excessive. However, when shallow foundations become impracticable a number of possibilities present themselves and each has to be considered and individually costed. For example, deep mass foundations, raft foundations or, as is frequently the case, piles would normally be considered.

It is clear that by creating the possibility of obtaining a favourable change in the limiting geotechnical conditions, ground improvement techniques have a profound effect on the traditional methods of foundation selection, especially as the result will be to allow the simplest forms of foundation to be retained on many weak soils or fills.

The feasibility of using the ground improvement approach in any given case depends on several factors, principally soil conditions, structure type, settlement tolerances and environmental considerations, and it is not universally applicable in, for example, the same way as piling. An elementary classification of the types of structure and ground condition favourable to the use of ground improvement techniques is shown in Table 1 where an attempt has been made to indicate the merit of ground improvement relative to other methods. As would be expected loose granular soils are especially amenable to improvement but in general terms the value of ground improvement rises as load density decreases and the size of the loaded area increases. A further factor not apparent from Table 1 is that in cases where depth of excavation varies widely the cost of ground improvement may be justified in that all foundations can be laid at a uniform level, avoiding the hazards inherent in excavation work and reducing construction time.

GROUND IMPROVEMENT BY VIBRO-COMPACTION

Vibro-compaction techniques are a well established means of stabilizing loose or weak soils and in a recent paper Greenwood (1970) has summarized the current state of the art in this field. The basic tool is a high powered poker-vibrator and this was originally developed as a means of achieving compaction in loose sands to any depth. The degree of compaction attained varies with grading and spacing of compaction centres but, in general, area treatment

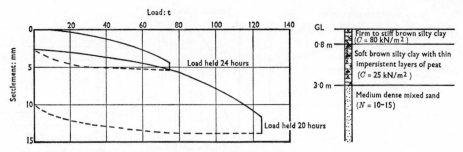

Fig. 2. Sewage treatment works; soil conditions and load test results

of sand involves a grid spacing of between 1·5 and 3 m, while the depth of compaction is made sufficient to absorb the main foundation stresses. In recent years European practice has seen the extension of vibro techniques to include treatment of many soft cohesive soils as well as a wide variety of fills. In clays the vibrator is used as a means of forming a hole through the soft material into which selected coarse aggregate is fed and compacted in stages as the vibrator is withdrawn. Although essentially unaffected by the action of the vibrator, the clay serves to confine the more rigid columns, thereby increasing the bearing capacity of the composite formation and reducing settlement. The terms vibro-replacement and vibro-flotation in British practice denote the use of compressed air or water as the jetting fluid employed on particular sites and do not affect the general principles involved (D'Appolonia, 1953; Grimes and Cantlay, 1965; Thorburn and MacVicar, 1968).

Test data illustrating the effectiveness of vibro-compaction in a granular soil are shown in Fig. 1. On this site, beneath a surface layer of fill, sand extended to a depth of about 21 m with sandstone below this. In the initial site investigation the SPT values in the sand indicated very loose conditions with N values of 3–6. Accordingly it was intended to use piles driven to bedrock for the proposed 14-storey office block and the foundation design consisted essentially of two parallel pile caps supported by a grid of 100 tonne capacity piles. On commencing piling it was found impracticable to drive isolated displacement piles through the sand to rock and thus work was suspended and additional boreholes carried out. A typical borehole is shown in Fig. 1 and it will be observed that the sand was in fact dense below about 9 m and generally medium-dense above. After consideration of the later borehole results, and in view of the difficulty likely to be achieved in driving piles to a sensibly uniform level through the medium-dense sand, it was decided to use vibro-flotation treatment to densify the sand down to 9 m. Treatment centres were arranged on a triangular pattern of side 1·7 m extending beyond the pile caps, this arrangement being designed to permit a safe bearing pressure of 430 kN/m² at shallow depth in the sand. In the treatment a granular infill of 5 mm to 20 mm was used as backfill at each treatment point and using vibro-flotation it was possible to stabilize the sand over the desired area to a uniform depth of 9 m. Kentledge loading tests were carried out both on the natural and compacted sand using bases 1·5 m square and results are also given in Fig. 1. In this instance the speed of the vibro-flotation operation minimized contract delay while the bearing capacity of the treated sand enabled the pile caps to be replaced by reinforced strip foundations of essentially the same dimensions.

Settlement control considerations are often the main criterion in foundation design especially where large loaded areas are located over alluvial clays and silts and this problem frequently arises in the construction of sewage purification plants. Vibro techniques were used on such a site in northwest England to stabilize an area of 200 m by 300 m beneath a series of filter beds where the imposed surface loading was generally 80 kN/m². The ground conditions

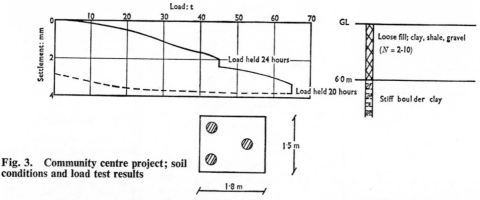

Fig. 3. Community centre project; soil conditions and load test results

(Fig. 2) consisted of soft clays with peat inclusions to a depth of approximately 3 m below working level with medium dense sands underneath. Excavation to site formation necessitated the removal of the original topsoil, and underlying thin desiccated crust, to the soft clays. To limit total and differential settlements over and between the various structures vibro-flotation was carried out on a general grid spacing of 2·3 m × 2·5 m, the stone columns penetrating through the soft alluvium and terminating in the sand. A number of kentledge load tests on 3 m square bases were carried out during the work and a typical result is shown in Fig. 2. It will be noted that although test load approached the ultimate bearing capacity of the natural clay, settlement on the treated ground was small.

Liquid storage tanks are clearly suited to the ground improvement approach when the soils, either due to shear strength or compressibility considerations, are inadequate for a normal flexible hardcore and bitumen foundation as the cost implications of adopting a pile and slab system are very high on this type of structure. Recently vibro-flotation was successfully used on a site where the ground conditions consisted of dredged clay fill and silt to a depth of approximately 7 m overlying medium-dense silty sand. Two tanks were involved, 55 m diameter by 18 m high, and of floating roof design. For each tank 7 m long stone columns were installed on a triangular pattern at centres of between 1·8 and 2·1 m over an area extending approximately 3 m beyond the tank perimeter. Under a water test, where a peak loading of 175 kN/m² was maintained for nine days, the maximum differential settlement across any diameter was less than 1 in 800 with correspondingly small differential movements around the perimeter of each tank. Apart from overall stability the principal design criteria for large tanks are differential settlements both across the tank and around the perimeter as these determine the risk of buckling in the tank wall or jamming of the floating roof in service. The differential settlement around the tank perimeter was within the tolerance of 1 in 360 specified by the tank manufacturer.

In older industrial areas the natural ground surface tends to be masked by a substantial thickness of unconsolidated fill or industrial debris and this feature often leads to foundation problems on new projects. Data from a typical problem are summarized in Fig. 3 and in this case a new community centre was to be constructed immediately next to existing swimming baths. Ground conditions consisted of colliery waste fill material to a depth of 6 m with stiff clays below this, and although the swimming baths were supported on piles, the consulting engineers responsible for the community centre also investigated the feasibility of using ground improvement. Their analysis showed that treatment of the fill by vibro-replacement produced substantial savings and this method was used to stabilize the full depth of fill beneath all the foundations. After treatment a design bearing pressure of 165 kN/m² was used at nominal

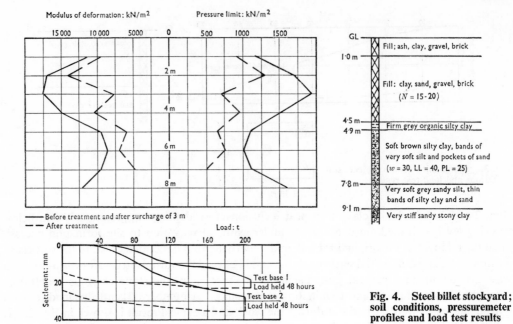

Fig. 4. Steel billet stockyard; soil conditions, pressuremeter profiles and load test results

depth and during the contract a kentledge load test was carried out on a working pad founda-tion 1·5 m by 1·8 m; the results are shown in Fig. 3 and it will be observed that negligible settlements were recorded under the test load of 230 kN/m².

GROUND IMPROVEMENT BY DYNAMIC CONSOLIDATION

The Ménard dynamic consolidation method uses controlled high energy surface tamping to compact weak soil. A key feature of this technique is its ability to engender compaction down to substantial depth, normally of the order of 12–15 m below ground surface and, perhaps of equal importance, experience has shown that careful phasing of the tamping passes and energy levels allows many low permeability silty and clayey soils to be compacted as well as virtually any granular material. The reduction in voids ratio causes strength and bearing capacity to rise while the post-treatment settlements are reduced and thus the benefits obtained using dynamic consolidation are similar to those produced by vibro treatment, but obtained without the addition of selected aggregates in the compaction operation. A detailed description of the dynamic consolidation technique may be found from the references, and in general it has been found to be a method especially suited to large open sites, open sites on reclaimed land, fills and alluvial soils (Ménard, 1972; Hansbo et al., 1973).

West and Slocombe (1973) described how the method was successfully used to compact a 10 m thick layer of granular colliery waste fill beneath a series of heavily loaded slipway founda-tions. Two further sites will now be discussed where dynamic consolidation was used to com-pact silty and clayey soils.

The first case concerns a covered storage area 125 m by 75 m where an average loading capacity of 250 kN/m² was required for the storage of steel billets; unlike many problems of this kind, the storage load was to be applied over the whole area and handled using overhead cranes. Thus there would be no reduction in loading at depth due to provision of access roadways within the storage area itself. The ground conditions are shown in Fig. 4. Rela-tive to the working level fill extended to a depth of approximately 4·5 m underlain by soft clays

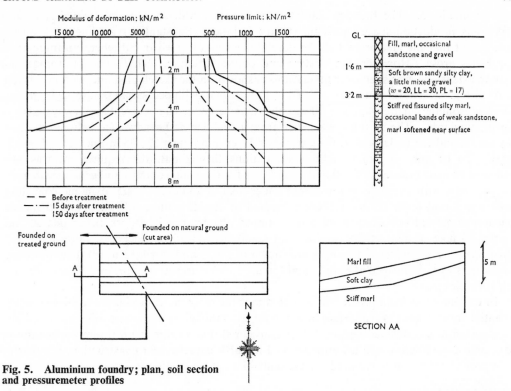

Modulus of deformation: kN/m²   Pressure limit: kN/m²

Fill, marl, occasional sandstone and gravel

Soft brown sandy silty clay, a little mixed gravel ($w = 20$, LL $= 30$, PL $= 17$)

Stiff red fissured silty marl, occasional bands of weak sandstone, marl softened near surface

– – Before treatment
– · — 15 days after treatment
—— 150 days after treatment

Founded on treated ground

Founded on natural ground (cut area)

Marl fill

Soft clay

Stiff marl

5 m

SECTION AA

N

Fig. 5. Aluminium foundry; plan, soil section and pressuremeter profiles

and silts containing occasional sand pockets with dense granular soils and stiff clays below depths of between 6 and 9 m.

The site investigation boreholes indicated that the undrained shear strength of the alluvium (30–40 kN/m²) was insufficient to provide an adequate safety factor under the storage loading and after examining the alternatives, dynamic consolidation was found to offer the cheapest and fastest method of overcoming the problem. The ground treatment was carried out in three passes using a 14 tonne tamper dropped with a free fall from heights of up to 18 m and typical pre- and post-treatment pressuremeter test profiles are shown in Fig. 4. It will be observed that, while the post-treatment limit pressure values (equivalent to ultimate bearing capacity) show that a safety factor of more than 3 was reached using dynamic consolidation, the pre-treatment values, while low, are nevertheless greater than anticipated from the site investigation. Although some of this deficiency could be ascribed to sampling and testing disturbance in the site investigation, the fact that an additional 3 m thick layer of fill material was placed over this part of the site in the interval between the site investigation and ground treatment may have led to a further significant increase in shear strength. However, it is important to note that the marked improvement shown in Fig. 4 was attained within a contract period of approximately eight working weeks using dynamic consolidation and, regardless of economic considerations, it is unlikely that further surcharging would have produced comparable results within this period. Apart from intensive pressuremeter testing, kentledge load tests were also performed on two 3·35 m by 2·15 m bases and the results are also shown in Fig. 4. It is of interest to note that excellent agreement was obtained between the measured deflexions and those calculated from the relevant pressuremeter readings.

Another application of dynamic consolidation in low permeability soil arose on the site for a new aluminium foundry and the soil conditions, location and pressuremeter test results are shown in Fig. 5. The foundry consists of a series of wide-span portal framed structures enclosing production space and ingot storage facilities. Substantial regrading work on the site meant that the greater part of the development was located in cut where no foundation problems arose. However, the western end of the structure, including ingot storage and future extensions, was sited over fill and this is shown on the sketch plan and section forming part of Fig. 5. The fill material consisted of Keuper marl with occasional sandstone fragments derived from the cut area, and this fill was placed and compacted to current motorway embankment standards. The fill thickness varied from 1 to 4·5 m and sandwiched between the base of the fill and the underlying stiff marls was a layer approximately 1·5 m thick of soft sandy clay. To be able to retain shallow foundations throughout, the requirement was permissible bearing pressures of 150 and 100 kN/m² for the main columns and general slab areas respectively, with a maximum settlement under load of 25 mm. Again, dynamic consolidation was found to be the most efficient method of attaining these tolerances and the treatment was carried out using a 14 tonne tamper, the treatment pattern being phased and varied to take account of the varying load requirements and fill thickness. Pre- and post-treatment pressuremeter test profiles are shown in Fig. 5 and it will be noted that after treatment tests were performed at both 15 and 150 days after completion of the treatment to observe thixotropic effects.

In the preceding examples it will be noted that the dynamic consolidation achieved a substantial improvement in the surface layers of fill, even though in both cases this was placed and compacted in a controlled fashion. It is considered that an important future application of dynamic consolidation will arise in the field of earth embankment construction, whether or not ground treatment is required for the embankment foundations.

FUTURE TRENDS

Specialized foundation techniques are in a constant state of evolution and progress is made principally as a result of experience gained in solving an ever increasing number of problems. There is a need for more research and field instrumentation aimed at the formulation of rigorous design methods for all real soil problems and a similar study is required for existing deep compaction techniques. So far as the general theme of ground improvement is concerned opportunities exist for further development particularly in connexion with the treatment of soft cohesive soils. Further advances may be possible in ground improvement methods not considered in this Paper, such as electro-osmosis, chemical injection and soil fracturing, and the ultimate objective should be a situation where all foundation problems can be solved by the ground improvement method.

REFERENCES

D'Appolonia, E. (1953). Loose sands—their compaction by vibro flotation. *American Society for Testing and Materials Special Technical Publication*, No. 156, 138–154.

Greenwood, D. A. (1970). Mechanical improvement of soils below ground surface. *Proc. Conf. Ground Engineering, ICE, London*, 11–22.

Grimes, A. S. & Cantlay, W. G. (1965). A twenty-storey office block in Nigeria, founded on loose sand. *Structural Engineer* 43, No. 2, 45–57.

Hansbo, S., Pramborg, B. & Nordin, P. O. (1973). The Vänern Terminal. An illustrative example of dynamic consolidation of hydraulically placed fill of organic silt and sand. *Sols-Soils*, No. 25, 5–11.

Ménard, L. (1972). The dynamic consolidation of recently placed fills and compressible soils. Application to maritime works. *Travaux*, No. 452.

Thorburn, S. & MacVicar, R. S. L. (1968). Soil stabilization employing surface and depth vibrators. *Structural Engineer* 46, No. 10, 309–316.

West, J. M. & Slocombe, B. C. (1973). Dynamic consolidation as an alternative foundation. *Ground Engineering* 6, No. 6, 52–54.

Compacting loess soils in the USSR

M. Yu. ABELEV*

The Paper deals with experience gained in compacting loess soils which collapse when saturated by water after being loaded by the weight of a structure or of the overlying soil. The installation of soil piles enables these soils to be compacted to a depth of 25 m before constructing foundations on them. After compaction the soils are no longer susceptible to subsidence when wetted. Moreover, they become stronger and less compressible. The Paper describes the application of soil piles in the USSR and the equipment used to sink the holes and to compact the local loess soils used to form the piles.

Le document décrit l'expérience acquise en compactant des loess qui s'écroulent après saturation d'eau, par suite du chargement du poids d'une structure ou d'une couche de sol supérieure. La mise en place de fondation pieux sol, permet le compactage de ces sols jusqu'à une profondeur de 25 m, avant la construction. Suite à la compaction, les sols ne sont plus susceptibles d'effondrement lors de leur saturation. D'ailleurs ils deviennent plus solides et moins compressibles. Le document décrit l'application des pieux de sol en URSS et l'équipement employé pour foncer les trous et pour compacter les loess locaux utilisés pour la formation des pieux.

Loess soils that subside on being wetted extend over vast regions in the USSR, USA, Romania, Hungary, China, India and many other countries. The depth of such subsiding soils may extend to hundreds of metres, and structures built on them are often damaged or destroyed unless special measures are taken. In the USSR this has led to development of construction methods on these soils. The specific principles followed in each case depend on the depth of potential subsidence, the characteristics of the subsidence and the types of structure to be erected.

In the USSR, the method most frequently employed for compacting shallow deposits up to 5 m deep is by means of heavy rammers. This method, proposed by the Author, is to drop a heavy rammer weighing from four to seven tons from a height of five to seven metres by means of a pile driver or crane. From ten to sixteen blows are applied to the same spot; this compacts the loose soil to a depth of 2·0–3·5 m.

More frequently soil piles are used for compacting deeper deposits of subsiding soils in areas where the thickness of these soils is more than 6 m. The piles may be up to 18 m deep. This method has been applied in the Soviet Union since 1948 and about 4·5 million cubic metres of loess soils have been compacted in this way.

METHOD OF INSTALLATION

Deep compaction by means of soil piles consists of the following operations:

(a) bringing the moisture content of the soil, wherever necessary, to the optimum value;

---

* Assistant Professor, Kuïbyshev Civil Engineering Institute, Moscow, USSR.

(b) sinking holes to the required depth in the foundation area by lateral displacement of the soil;

(c) filling the holes thus obtained with soil by layer-by-layer compaction to the required density.

The use of tower pile drivers enables a number of holes to be drilled from a single position by means of a single-acting steam-and-air hammer weighing at least three or four tons and a drive pipe. The latter is seamless steel tube with an outside diameter of 280 mm and a wall thickness of at least 16 mm. Since the tubes are manufactured in lengths of 6–8 m, the drive pipe is made of several lengths joined together by outside or inside couplings or by welding. In driving, the tube is placed on a conical shoe with a diameter 1·5 times that of the tube. The resulting enlarged hole enables the tube and hammer to be lifted out of the ground by means of the pile driver hoist. Tips for sinking holes have a conical striking point with an included angle of 30° and a cylindrical element 150–200 mm high. In sinking holes, a considerable enlargement of their top diameter is observed. This can be prevented by installing a conductor tube at the surface before sinking the hole; such a conductor is a cylinder of plate steel, 55 cm in diameter and 75 cm high.

A pile driver can compact the soil to a depth of 14–16 m, taking from 16 to 42 minutes (270 to 486 blows of a pile hammer weighing 3 tonnes) to sink one hole. Experience shows that it is advisable to sink alternate holes at the first stage, the second set being then sunk after filling the first set with soil material. This avoids caving due to sinking holes very close to each other.

INSTALLATION USING BLASTING

The method of deep compaction of subsiding soils by blasting consists of drilling a blasthole, 75–80 mm in diameter, to the required depth of compaction. Then a charge of explosives, consisting of separate cartridges joined together in a chain, is lowered into the hole. The diameter of the explosive cartridge is one tenth of the design diameter of the hole produced by the blasting. Instantaneous detonating fuse type DShA, with a rate of detonation of 6500 m/s, is attached to the cartridges.

It has been established experimentally that for macroporous loess soil of optimum moisture content, in which the strength and deformational properties are essentially determined by the clay fraction, the required number of cartridges (42–45 mm in diameter and weighing 50 g) per running metre of blasthole can be related to the plasticity index.

At sites where the soil is dry the compacting effect is ensured by raising the moisture content to the optimum value by adding water, the amount required being determined on the basis of the coefficient of permeability and rate of capillary movement of water in the vertical and horizontal directions. The process of raising the moisture content is accelerated by injecting water through the holes of a depth 0·7 to 0·8 of that in which the moisture content of the soil is to be raised. At depths over 6 m, wetting is carried out through a network of holes produced by wash boring. The bottom of the foundation pit and the holes are filled with a porous material such as fine-graded broken brick. Raising the moisture content to the optimum value is of especial importance when using blasting methods of compaction; if the soil is too dry and stiff, scaling of the overcompacted soil crust and caving in of the hole is observed. Caving in, according to experimental data, does not exceed one eighth of the hole depth when the soil is at optimum moisture content.

The blasthole for lowering the charge should be from 60 to 80 mm in diameter, which is sufficient to obtain the required air gap between the cartridge and hole, and to permit a charge of cartridges 40 to 50 mm in diameter to be lowered freely.   In dry soils requiring preliminary moistening, such blastholes are made by vibratory drilling; in other cases, blastholes may be obtained by driving a drilling rod with a tip diameter of 80 mm.   The length of the explosive charge is equal to the depth of compaction.   The uppermost cartridge is located 50 to 70 cm below the top of the blasthole.   The charge is hung from a wooden batten and securred to cords which are tied to the cartridges.   The charges are detonated separately in each blasthole with an interval between blasts of at least 1 min.   A simultaneous blast in several holes is not permitted because the total energy of the explosion may sometimes spread beyond the limits of the part of the foundation base to be compacted.   After blasting and discharge of the gases, the hole is plumbed for depth, and in the first holes to be installed the diameter and its variation with depth are also measured

In experimental work, deep compaction was accomplished sinking holes 50 cm in diameter. This enabled the number of soil piles in the compacted foundation base to be reduced, and led to a general increase in labour productivity in deep soil compaction.   The new technique found immediate application in construction practice and was employed for the deep compaction of soils to depths of 8 and 16 m in the foundation bases of various structures.

Cable, or churn, drilling is done by a string of tools operating with a drop of 1·2 m at a frequency of 40 to 50 blows per minute.

FILLING THE HOLES

The holes obtained by one of the methods are filled with local loess or other loams and sandy loams, not containing construction debris or coarse materials.   Rammer tips used in compacting the soil when filling the hole are designed in the form of a paraboloidal wedge.   Tests of these tips have shown their efficiency in sinking and compacting.   The tip diameter of 42·6 cm compacts the soil to form holes from 50 to 55 cm in diameter due to the swinging of the string of tools.   The filling soil should be compacted to the maximum density for the type of soil being used.   For loams with typical moisture content values of 14–18% at the plastic limit, the soil density (unit weight of dry soil) for the most highly compacted structure, taking the entrapped air into account, ranges from 1·7 to 1·8 tonnes per cubic metre.

Special investigations were conducted to determine the amount of work required for compaction.   The soil was placed in batches of 100 to 200 kg into holes 40 to 45 cm in diameter obtained by blasting, and then compacted by rammers weighing from 350 to 450 kg.   About seven tonne-metres of work must be done to compact 100 kg of loose soil of optimum content to a density of 1·7–1·75 t/m³.

The hole is filled with soil in batches so that the column of soil material does not exceed 1·2 m in height.   The required degree of compaction is obtained if the height of the column of soil material loaded into the hole in each batch does not exceed two hole diameters.

A new technique is to compact each batch of soil material by a cable (churn) drilling rig having a string of tamping tools weighing at least 1 ton.   The number of blows for compaction is calculated on the basis of doing at least 10 tonne-metres of work to compact each 100 kg of soil in the batch.   The drilling rig compacts a batch of soil weighing from 250 to 300 kg in 25 blows.

The soil piles are filled in and compacted after checking the depth of hole obtained by sinking or blasting.   If the soil was compacted by blasting, the height of the column of soil fallen from the walls is measured before it is compacted.   Before starting to fill the hole, it is necessary to compact the soil fallen from the walls.   To reach the specified density, the fallen soil, usually overcompacted by the blast, is moistened by a loam slurry having a moisture content near to the liquid limit. This slurry adheres to the lumps of fallen soil and raises their moisture content to the optimum value. The fallen soil is compacted for a half an hour after pouring in the slurry after which the hole is filled in with further soil at optimum moisture content. In deep compaction with pile driving equipment or cable drilling rigs, no accumulations of fallen soil were observed.

To overcome the difficulty of locating the soil piles in the compacted foundation base, the last batch of filling material, placed in the hole at the foundation level, is mixed with a small amount of black binding material, such as bitumen or tar, or with humus-enriched soil.

REFERENCES

Abelev, Yu. M. (1936).   Making soil piles in macroporous loess soils by blasting.   *The building industry*, No. 4.   Moscow (in Russian).

Abelev, Yu. M. (1939).   Main results obtained in investigating features of the constructional properties of loess soils, and methods for their strengthening.   Papers on Construction on loess soils.   *Ukr. NITO Stroit*.   Kiev (in Russian).

Abelev, Yu. M. (1946).   Influence of the pile material on the behaviour of a floating pile foundation.   *Construction engineering bulletin*, No. 17–18.   Moscow (in Russian).

Abelev, Yu. M. & Abelev, M. Yu. (1968).   Fundamentals of design and construction on macroporous subsiding soils.   Moscow: Stroiizdat Publishers (in Russian).

Abelev, Yu. M. & Galitsky, V. G. (1962).   A new machine designed by the Research Institution for Foundation Bases for ramming soil piles.   *Proc. Conf. Soil Stabilization and Compaction, Kiev* (in Russian).

Buryak, P. G. (1961).   Experience in the design and construction of the buildings and structures of coke and by-products plants on loess soils.   *Proc. Conf. Construction on Loess Soils*.   Kiev: Ukrainian Academy of Construction and Architecture Publishers (in Russian).

Gubanov, E. K. (1961).   Experience in the depth compaction of loess soils in Krasnoyarsk—problems of construction on loess soils.   *Proc. Inter-Institute Scientific Conf., Voronezh* (in Russian).

Building structures supported by stabilized ground

S. THORBURN, FICE, FIStructE, FASCE, FGS*

The Paper introduces the use of the concept of stabilized or reinforced ground as a means of limiting the long-term differential movements of a building. The techniques of strengthening superficial and artificial deposits utilizing depth vibrators are described together with the response of ground to treatment by vibro-compaction and vibro-replacement. The Paper also defines the semi-empirical approaches to design derived by the Author from field experience.

La communication présente le principe de l'utilisation du sol stabilisé ou renforcé afin de limiter les mouvements différentiels à long-terme, produits dans un immeuble. La technique de renforcement des dépôts artificiels et superficiels par l'utilisation des vibrateurs profonds, est décrite en même temps que la réaction du sol au traitement par vibro-compactage et par vibro-remplacement. En plus, on décrit les approches semi-empiriques proposées par l'auteur à la suite de ses expériences in situ pour l'étude des projets de fondations.

The concept of stabilized ground is introduced by considering a structure constructed on a concrete slab foundation supported by a large number of friction piles. If the slab foundation is capable of directly transmitting a significant proportion of the total vertical loads to the superficial deposits within the large pile group, then the building structure is virtually supported by a block of reinforced soil. Whereas, in normal circumstances, the rigidity of a building structure can be the dominant factor in the behaviour of the building, the introduction of stabilized ground beneath a structure also plays an important part in structural behaviour. If the structure is able to tolerate some differential settlement then an economic balance may be achieved by providing a block of stabilized ground that has adequate stiffness to limit the long-term differential movements of the structure.

The strengthening of the ground may be accomplished by compaction if the ground is non-cohesive, or by providing reinforcement in the form of uncemented columns of aggregate. Piled foundations need not be a first consideration. The techniques of strengthening superficial and artificial deposits by means of depth vibrators has been developed within the UK over the past twelve years. In the Glasgow district the concept of stabilized ground using stone columns was first applied in the autumn of 1962 (Thorburn and MacVicar, 1974), when a long continuous three-storey block of flats was founded on ground strengthened by means of depth vibrators. However, the concept of stabilized ground, although readily defined, is not completely understood because of the complex nature of the problem. As a result the depth vibrator techniques of strengthening ground have been developed largely on the basis of field tests and experience in combination with a thorough knowledge of soil mechanics.

The formation of blocks of stabilized ground utilizing depth vibrator techniques have resulted in satisfactory building performances but it cannot be concluded on this evidence that a complete understanding exists of the concept. This Paper describes the response of ground

---

* Thorburn & Partners, Glasgow, Scotland.

to treatment by depth vibrators, defines the semi-empirical approaches to design developed by the Author on the basis of his experience, and indicates the probable reaction of ground to applied building loads.

Cost comparisons have been made by Basore and Boitano (1969), Watson and Thorburn (1966) and Penman and Watson (1967); their conclusions are given in Table 1.

STRENGTHENING OF SUPERFICIAL DEPOSITS

Response of ground to treatment

Vibro-compaction. The characteristics and assembly of a typical depth vibrator have been covered by Greenwood (1970) and Mitchell (1970) and will not be discussed in this Paper. It is sufficient to state that rotating eccentric weights within the depth vibrator impart to the body of the vibrator a gyratory motion in a horizontal plane. When a depth vibrator is introduced into a cohesionless soil the particles are agitated by the horizontal vibrations caused by the gyratory motion of the vibrator. Research has indicated that the vertical components of the forced vibrations are unimportant. The frequency of the soil vibrations is the same as the frequency of the vibrator which is equivalent to the speed of the motor which rotates the eccentric weights. The maximum amplitude is obtained when the vibrator is freely suspended in air. The amplitude of the vibrations in the soil is dependent on the power of the vibrator and if the maximum amplitude of the vibrator can be maintained during the compaction process, then adequate power is available.

When the vibrator penetrates a cohesionless soil the intergranular forces between the soil particles are temporarily nullified and liquefaction occurs in the immediate vicinity of the vibrator within distances of about 300 to 550 mm from the outer vibrating surface of the body of the vibrator. Beyond these distances liquefaction is incomplete because of damping effects. As the vibrator continues to penetrate the cohesionless soil the forces of resistance to penetration increase and the power demand to ensure continuous penetration also increases. If the power of the vibrator is adequate then the maximum amplitude is maintained but if the available power is smaller than that required then the amplitude is reduced in proportion to the power available. The soil liquefaction induced by the horizontal vibrations permits the soil particles to be rearranged without stress by gravitational forces and since the particles are unstressed, no release of stress can take place after compaction and, therefore, the cohesionless soil is permanently strengthened.

Cohesionless soils compacted by depth vibrators will adequately resist changes in compaction by dynamic stresses induced by external sources of vibration since these stresses may generally be expected to be less than those induced by the intense vibrations experienced by the soil during the compaction process. The compressibilities of cohesionless soils are considerably decreased by the compaction process and the angles of shearing resistance are increased.

Table 1

Ground conditions	Alternative foundation	Cost of stabilized ground
		Cost of alternative foundation
9·1 metres of loose sand	Compaction piles	0·76
Pulverized fuel ash hydraulically deposited on peat and silty clay	RC piled foundations	0·44
Soft laminated silty clay fill material hydraulically deposited on sands	RC piled foundations	0·68

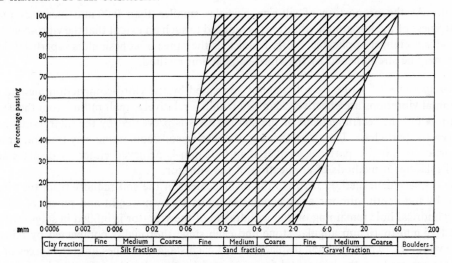

Fig. 1. Range of soils which may be strengthened by vibro-compaction

An adequately powered depth vibrator is quite capable of penetrating cohesionless soils under its own weight and power of vibration but often water under pressure is emitted from the conical nose of the vibrator during penetration.   The water is primarily provided to maintain the stability of the sides of the hole formed by the depth vibrator during the compaction process.   The diameter of the cylindrical mass of soil effectively influenced by the horizontal vibrations depends on such factors as the particle size distribution of the soil, the presence of groundwater, and the amplitude and frequency of the vibrator.

No significant compaction takes place beyond a distance of 2·50 m from the vibration centre regardless of the length of time the vibrations are maintained.   By suitable spacings and depths of penetration of the individual vibration centres, soil masses of the required dimensions can be relatively uniformly compacted.   Experience has shown that the density of a cylindrical mass of soil at the vibration centre having a diameter approximately equivalent to that of the vibrator can be less than the density of the compacted soil surrounding the vibration centre.   This phenomenon may be attributed to soil arching effects.   For practical design purposes, however, this effect may be ignored.   The spacings of the vibration centres in fine-grained cohesionless soils must be closer than in coarse-grained cohesionless soils, because of the greater damping effects of the fine soil on the induced horizontal vibrations.

With regard to the time of vibration most of the compaction takes place within the first two to five minutes at any given elevation of the vibrating body within the cohesionless soil and it is uneconomical to attempt to achieve a denser state of compaction by excessively increasing the time of vibration.   It should be appreciated that the source of vibration within the depth vibrator is located within the first three metres from the conical nose of the vibrator and that transmission of vibration to the upper extension tubes, to which the body of the vibrator is attached, is virtually eliminated by the provision of special isolator couplings.

The preceding paragraphs have essentially covered the response of silt-free sands and gravels to the horizontal vibrations induced by depth vibrators but as the proportion of silt-size particles increases then damping effects become so significant that the radius of compaction around the vibration centre is greatly reduced.   Experience has established that soil containing up to 30% of silt particles having particle sizes within the range 0·02 to 0·06 mm can be effectively compacted by the horizontal vibrations provided the centres of vibration are closely spaced.   As the proportion of silt-size particles smaller than 0·06 mm increases then the effectiveness of the horizontal vibrations is reduced to such an extent that the vibro-compaction method is no longer applicable.   Fig. 1 indicates the range of cohesionless superficial deposits which may be strengthened by the vibro-compaction method.

Vibro-replacement.   Cohesive soils cannot be treated by the vibro-compaction method as the horizontal vibrations are effectively damped within a relatively small radius from the centre of vibration and the rearrangement of the soil particles is prevented by the intergranular forces between the particles.   The strengthening of cohesive soils containing clay fraction or a large proportion of silt-size particles smaller than 0·02 mm is achieved either by displacement or non-displacement methods of application of the vibrator.

The displacement method involves the displacement of the soil radially from the vibration centre by the depth vibrator as it penetrates under its own weight and power of vibration. The cylindrical hole remaining on the withdrawal of the vibrator is infilled in stages with well-graded 75 mm to 10 mm angular stone and each stage thoroughly compacted by reinsertion of the vibrator.   The radial soil displacement at each stage continues until the forces of resistance are greater than the horizontal forces exerted by the vibrator.   The mode of formation of the generally irregular, roughly cylindrical shaped stone column ensures that the passive resistance of the surrounding cohesive soil is fully mobilized at very small horizontal radial strains when the stone column is subjected to vertical loading.

The displacement method is of advantage when strengthening partially saturated soils since the radial stresses induced by the soil displacement result in a significant gain in soil shear strength between the centres of vibration.   The displacement method is also used to strengthen fully saturated cohesive soils within the range of undrained shear strength of 20 to 50 kN/m² for urban sites where the use of water for jetting purposes would not be permitted for environmental reasons.

Although no jetting water is used with the displacement method it is essential to use compressed air to break the suction which develops when the vibrator is withdrawn from a soft saturated cohesive soil, otherwise the sides of the hole collapse and difficulty is experienced in forming a satisfactory stone column.   Great care must be taken when free water is present within the hole as the disturbance caused by the emission of compressed air can create a soft clay slurry which will contaminate the stone column.   Indeed, contamination of the clean stone fill material must be avoided at all times since the shearing resistance is significantly reduced if clay slurry is permitted to smear the interfaces between the stones.   Concomitantly, the strength of a stone column is greater if angular stone fill material rather than rounded gravel fill material is used to form the column.

There are major difficulties in forming uncontaminated stone columns by the displacement method in very soft soils because the radial displacement seriously distorts and remoulds the surrounding soil inducing excess pore-water pressures which cannot dissipate rapidly even in a laminated soil because the distortions destroy the natural horizontal drainage laminae.   When the building loads are applied the rate of gain of shear strength of the very soft soil between and surrounding the stone columns is also significantly reduced.   The initial reduction in the shear strength of the soft silty clays of the Grangemouth district due to the remoulding effects of the displacement method of vibro-replacement is clearly shown in Fig. 2.   The non-displacement method of vibro-replacement must be used in very soft saturated cohesive soils having undrained shear strengths less than 20 kN/m², and this method involves the removal of soil by water emitted under pressure through jetting holes in the conical nose of the vibrator until a cylindrical hole of the required diameter is formed.

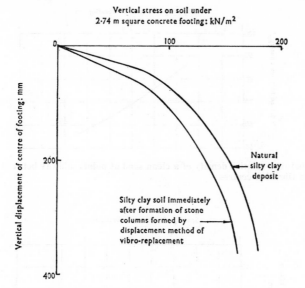

Fig. 2. **Results of load tests on natural and treated clay soil of the Grangemouth district**

The Author has never used the non-displacement method of vibro-replacement for building structures where soils were encountered having undrained shear strengths less than 14 kN/m² because of the relatively low radial support given to the stone columns by the soils. Johann Keller GmbH has overcome the problem of using stone columns in such soils by mechanically coupling three or four depth vibrators together which permits the very soft cohesive soils to be removed by jetting over a plan area of four square metres resulting in the formation of unusually large stone columns.

The vibro-replacement method of strengthening ground is not recommended for deep deposits of highly organic silts and clays or peats. If these soils have been intercalated with sand deposits in layer thicknesses not exceeding about 450 to 600 mm, then it is possible to re-place effectively and economically these thin layers by the compact stone fill materials which form the stone columns. Unconfined peat layers such as basin bog material cannot be treated by vibro-replacement methods. Thick layers of confined peat and organic very silty clays have been effectively strengthened and reference may be made to papers by Thorburn and MacVicar (1974) and Watson and Thorburn (1966).

The decision to utilize vibro-replacement methods of strengthening very soft cohesive soils having shear strengths less than 20 kN/m², confined layers of organic silts and clays, and peat, should be made only after careful analyses of soil data obtained from an adequate site investi-gation and economic comparison with other types of foundation.

Empirical design approach

Vibro-compaction. On the basis that the depth vibrator has adequate power to overcome the forces of resistance and maintain its maximum amplitude then the increase in the relative density of a cohesionless soil due to the horizontal forces of vibration depends on the spacing of the centres of vibration. Fig. 3 gives an indication of the relationship between the relative density of a clean sand at points midway between centres of vibration formed in a triangular

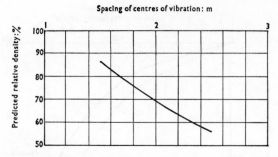

Fig. 3. Relationship between relative density of a clean sand at points midway between the centres of vibration and the spacings of the vibration centres

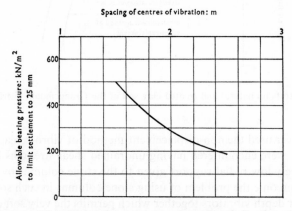

Fig. 4. Relationship between allowable bearing pressure and spacing of vibration centres for footings having widths varying from one to three metres, founded on cohesionless soil

pattern, and the spacings of the vibration centres. It should be appreciated that the relation-ship depends on the particle size distribution of the soil, the presence of a water table, and the amplitude and frequency of the vibrator.

The relative densities of cohesionless soils strengthened by vibro-compaction will vary in a non-linear manner from about 100% at distances of about 300 to 550 mm from the centre of vibration to the approximate values given in Fig. 3 at points midway between the vibration centres. This non-uniformity of compaction may be neglected for design purposes, and Fig. 4 indicates the allowable bearing pressures which may be expected from cohesionless soils for various spacings of vibration centres. The allowable bearing pressure for shallow spread footings is generally controlled by settlement considerations rather than by bearing capacity and the information given in Fig. 4 may be used for preliminary design purposes for footings having widths varying from one to three metres, founded on a cohesionless soil strengthened by vibro-compaction. It should be appreciated that the response of cohesionless soils to treatment will vary from site to site and therefore, that Fig. 4 is only given as an aid to design. *Vibro-replacement.* The strengthening of cohesive soils is entirely dependent on the formation of very dense columns of coarse angular stone since no compaction is achieved by the hori-zontal forces of vibration. The presence of the stone columns generally results in improve-ments in the rate of dissipation of excess pore-water pressures and the rate of gain in shear strength due to improved radial drainage conditions, but these beneficial effects are generally

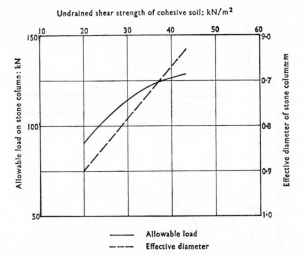

Fig. 5. Relationship between allowable working load on stone column and undrained shear strength of cohesive soil at point of maximum radial resistance

ignored in design since, in fully saturated soft cohesive soils, there is often an initial reduction in shear strength due to the shear displacements causing remoulding.

The design approach currently employed by the Author for building structures is that recommended by Thorburn and MacVicar (1968), whereby the total building loads are supported entirely by the stone columns. Such an approach ensures adequate factors of safety in respect of the bearing capacity of the stabilized ground and provides ground of considerable stiffness. The ultimate bearing capacity and the stiffness of the stabilized ground can be varied by permitting a proportion of the building loads to be supported by the soil between the stone columns but theoretical solutions for the prediction of the strength and stiffness of stabilized ground have yet to be established.

Figure 5 shows the relationship between the allowable working load to be used for preliminary design purposes and the undrained shear strength of the cohesive soil at the depth of maximum radial resistance (Hughes and Withers, 1974), within the normal range of undrained shear strengths of cohesive soils which can be strengthened by the displacement and non-displacement methods of vibro-replacement.

The non-linear relationship given in Fig. 5 was obtained from consideration of Rankine's theory of passive soil pressures modified for radial strain and from field measurements of average diameters of stone columns relative to the undrained shear strengths of the cohesive soils within which the stone columns were formed. However, the field measurements of column diameter used in the development of Fig. 5, are related to stone columns formed by the powerful Cementation and Keller vibrators and it would be prudent to establish the diameter of stone columns for particular models of vibrators and particular ground conditions.

Figure 6 shows the allowable stress on a stone column adopted for the development of Fig. 5, together with the allowable stress on a stone column derived by Hughes and Withers (1974), who have demonstrated that, for soils having uniform shear strength with depth, the base and bulging failure of a stone column occur simultaneously when the column length is four times the column diameter.

It is recommended that major vibro-replacement works should be controlled by random measurements of the diameters of the stone columns together with load tests to failure on

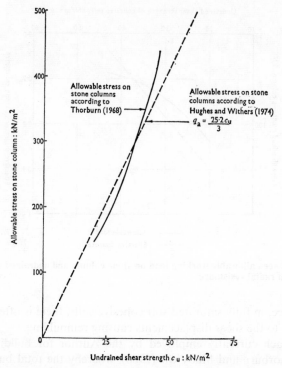

Fig. 6. Relationship between allowable vertical stress on stone column and undrained shear strength

representative stone columns. It should be appreciated that a load test on a single stone column is a measure of workmanship together with the capability of the vibrator to form stone columns of adequate diameters and not a measure of the ultimate bearing capacity of the stabilized ground. The condition of randomly selected stone columns should also be inspected after exposure by excavation in order to ensure the formation of a very dense column of clean stone.

Reaction of ground to applied building loads

Vibro-compaction. The structure of cohesionless soils is affected in two ways by vibro-compaction. In its simplest form, where no coarse fill material is introduced into the ground at the centre of vibration, the cohesionless soil is thoroughly compacted by the horizontal forces of vibration to form a reliable foundation material, which is uniformly very dense within the depth of strengthening. Any horizontal variations in relative density between the vibration centres may be neglected for practical purposes. In fine cohesionless soils, however, difficulty can be experienced in compacting the soils purely by the horizontal forces of vibration and coarse fill material must be introduced into the water-filled annulus around the depth vibrator to form columns of dense coarse gravel within the fine-grained cohesionless soils. Although a relatively more complex soil structure is formed by the presence of the vertical columns of gravel than in the simplest form of vibro-compaction, the reaction of the ground to applied building loads may be considered for most engineering purposes to be equivalent to that of a cohesionless soil in a very dense state of compaction. In special circumstances where the building loads are heavy, settlement is critical, and the centres of vibration are closely spaced, the engineering properties of the combined soil structure of dense coarse vertical columns surrounded by dense fine cohesionless soil may require to be considered in the foundation analyses.

Vibro-replacement.   Regardless of whether the displacement or non-displacement method of vibro-replacement is used to strengthen cohesive soils the resulting ground structure is complex and the reaction of such stabilized ground to applied building loads is quite different from that of cohesionless soils strengthened by vibro-compaction.

When building loads are transmitted by concrete footings to the surface of cohesive soils reinforced by stone columns, a large proportion of the total load is initially resisted by the very dense stone columns which are rigid relative to the surrounding soft cohesive soils.   The remainder of the building load is carried by the soft cohesive soil in contact with the concrete footings.   The initial loads on the tops of the stone columns produce radial strains in the cohesive soils surrounding the roughly cylindrical walls of the stone columns thereby mobilizing the radial resistance of the soil.   The magnitudes of the radial strains necessary to develop significant radial resistances are small because of the considerable radial preloading that occurred during the formation of the stone columns.   The vertical strains in the stone columns produced by the initial application of the building loads cause a transference of load from the yielding column to the soil under the concrete footing.   As consolidation of the soil takes place load is transferred from the yielding soil to the tops of the relatively rigid stone columns which then experience further vertical strain, and so on, until an equilibrium condition is reached.

Although the response of soft cohesive soil strengthened by dense stone columns is understood qualitatively, the complexity of the soil–column interaction problem does not permit a simple solution.

Hughes and Withers (1974) have recommended a solution which they consider produces an upper bound to the estimate of settlement by considering compatibility of vertical strains between the dense granular columns and the surrounding soft cohesive soil.   Such an approach gives an indication of the vertical ground displacements provided that the stress–strain relationships for both the column and soil are known.   The effects of consolidation on the stress–strain relationships must be taken into consideration in such an approach and the short-term relationships which would be obtained from relatively rapid plate loading tests are not applicable.   However, for most building structures, for which settlement is an important consideration, it is recommended that the total building loads should be supported entirely by the stone columns.

Such an approach will limit the magnitude of total settlement which will be experienced by building structure and permits the prediction of settlement to be made in a relatively simple manner.   The total settlement may then be closely approximated by the final vertical strain at the tops of the stone columns due to the stresses imposed by the total building loads plus the compression of the soils which exist beneath the layer of ground strengthened by vibro-replacement and which are significantly stressed by the concrete footings.

The research work by Hughes and Withers (1974) indicates that the vertical displacement of the tops of the stone column within the range of working stresses is less than half the maximum radial strain in the column.   As previously indicated, the magnitudes of radial strain necessary to develop working stresses are small because of the considerable radial displacements of the cohesive soil which occur during the formation of the stone column.   Field tests have demonstrated that the vertical displacement of the top of a densely packed clean stone column formed by a depth vibrator of adequate power at the working loads given in Fig. 5, may be expected to range from 5 to 9 mm.   A conventional soil mechanics approach will provide a prediction of the settlement of the soils beneath the layer of reinforced ground to which should be added a value of 5–9 mm to allow for the vertical compression of the layer strengthened by the formation of the stone columns.

Economies can be achieved by permitting some proportion of the total building loads to be supported by the soft cohesive soils under the concrete footing and between the columns but the prediction of total settlement presents analytical difficulties and further full-scale field research is required to permit such a solution to be adopted with complete confidence for important building structures.

It is worthy of note that the research work of Hughes and Withers (1974) indicates that the ultimate stress in the stone columns was obtained only when the vertical displacement of the column was 58% of the column diameter. Field test results examined by the Author on full-size stone columns indicate that the vertical displacements of the tops of the stone columns at failure, were only of the order of 10–15% of the diameters of the columns.

STRENGTHENING OF ARTIFICIAL DEPOSITS

Response of ground to treatment

Essentially cohesionless fill materials. Randomly deposited heterogeneous fill materials which have been subjected only to the weight of vehicular traffic which transported the materials to the site and which consist basically of essentially cohesionless materials such as boiler ash, or finely fragmented shale from old mine workings, are not affected by the horizontal forces of vibration to the same extent as clean cohesionless soil, such as sand. The strengthening of such materials is achieved chiefly by the radial forces of displacement caused by the formation of stone columns at the centres of vibration. The displacement method of vibro-replacement is generally the most effective means of strengthening essentially cohesionless made ground.

The degree of improvement in the load bearing properties of made ground depends on the composition and disposition of the heterogeneous fill materials which can never be accurately determined from a normal site investigation. The information provided by the site investigation boreholes together with field penetration or pressuremeter tests, should always be supplemented by visual examination of the fill materials exposed by inspection pits sunk at selected positions within the site for development, and, whenever possible, the source of the fill materials should be determined from local records. Experience has demonstrated that made ground containing large pockets of household refuse, bales of textile fibre, or masses of vegetable matter cannot be reliably strengthened by either vibro-replacement or vibro-compaction because the disposition and extent of these most unsuitable materials cannot be established. The radial strains experienced by the stone columns due to the presence of these highly compressible waste materials can cause significant non-uniform vertical foundation displacements.

Made ground containing heavy demolition spoil such as old concrete lintels which form rough arches resulting in cavities within the essentially cohesionless made ground can present considerable difficulties in the application of the vibro-replacement process; similarly, the existence of an excessive quantity of old timber planks or doors within otherwise granular demolition spoil can negate the effectiveness of the process. Pulverized fuel ash can be effectively strengthened by vibro-compaction or by the displacement method of vibro-replacement and reference should be made to the field tests by Watson and Thorburn (1966).

Essentially cohesive fill materials. Cohesive fill materials are often partially saturated and are readily strengthened by the displacement method of vibro-replacement. Artificial deposits consisting essentially of cohesive fill materials respond to treatment in a manner similar to

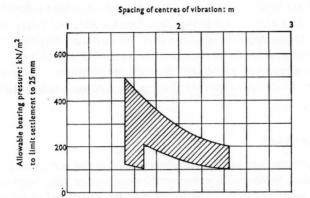

Fig. 7.   Relationship between allowable bearing pressure and spacing of vibration centres for footings having widths varying from one to three metres founded on made ground

that of partially saturated cohesive superficial deposits.   Care should, however, be taken if groundwater is present to prevent contamination of the stone columns by the penetration of soft wet clay into the stone column.

Empirical design approach

Essentially cohesionless fill materials.   If the artificial deposits consist of coarse fill materials which are cohesionless and completely responsive to the horizontal forces of vibration, then, for preliminary design purposes, the upper limit in Fig. 7 may be used.   However, it is normal experience to find that even mainly granular fill materials contain fine fraction which imparts cohesive properties to the artificial deposits and, therefore, the horizontal forces of vibration are less effective.   The heterogeneity of made ground together with the problem of adequately defining the variations in the engineering properties of this ground render it impossible to present a unique curve which relates allowable bearing pressures to the spacing of the centres of vibration.   Experience has indicated that one may expect a family of curves for different sites which may be expected to lie within the hatched zone given in Fig. 7.   Only engineering judgement will enable the centres of vibration to be selected for design purposes as a normal site investigation presents limited information and even plate loading tests provide insufficient information to enable the variations in the engineering properties to be evaluated.

Essentially cohesive artificial deposits.   Figs 5 and 6 may be used for preliminary design purposes but care should be taken when assessing the undrained shear strength of the cohesive fill materials to be used in conjunction with Figs 5 and 6 as clay fill materials were often deposited in the form of lumps of clay which soften non-uniformly.   Once again, engineering judgement is required and examination of the exposed surfaces of artificial clay layers within inspection pits is strongly recommended as an aid to judgement.

Reaction of ground to applied building loads

Essentially cohesionless artificial deposits.   The reaction of essentially granular made ground to applied building loads is similar to that described earlier in the section on strengthening of superficial deposits.

Essentially cohesive artificial deposits. The reaction of essentially cohesive made ground to applied building loads is also similar to that described for superficial deposits with the advantage that cohesive artificial deposits are often partially saturated resulting in the generation of relatively lower excess pore-water pressures. The rate of dissipation of pore-water pressure may be expected to be relatively rapid and, therefore, equilibrium of the complex soil–column structure is achieved much more quickly than in the case of a fully saturated soft cohesive superficial deposit.

CONCLUDING REMARKS

The object of this Paper is to present the observations of the Author over a period of twelve years, of an important geotechnical process; the conclusions have been deduced from field experience and not from research. The Paper may be considered as an attempt to provide data for guidance in making reliable but economical engineering decisions and not as a research document.

The depth vibrator provides engineers with the means of strengthening ground to controlled shallow depths. Although the response of artificial and superficial deposits to the horizontal forces of vibration and radial displacement is known, further research is required to establish design criteria for different ground conditions. Too few full-scale load tests on concrete footings have been carried out for engineers to have a complete understanding of the problem. Load tests on individual stone columns or centres of vibration do not provide the necessary information, since the problem is analogous to the difference in behaviour between single piles and groups of piles.

ACKNOWLEDGEMENTS

The Author is indebted to the many UK and other European engineers who have discussed with him at various times the subject matter of this Paper but wishes to acknowledge, in particular, the guidance of his mentor, Dipl Ing Carl Rappert, and the constructive criticism and advice given by Dr T. Whitaker and Dr J. Burland of the Building Research Establishment, England.

REFERENCES

Basore, C. E. & Boitano, J. D. (1969). Sand densification by piles and vibroflotation. *Jnl Soil Mech. Fdn Div. Am. Soc. Civ. Engrs*, SM6, 1303–1323.

Greenwood, D. A. (1970). Mechanical improvements of soils below ground surface. *Ground Engineering Conference, ICE*, 11–22.

Hughes, J. M. O. & Withers, N. J. (1974). Reinforcing of soft cohesive soils with stone columns. *Ground Engineering*, 42–49.

Mitchell, J. K. (1970). In-place treatment of foundation soils. *Jnl Soil Mech. Fdn Div. Am. Soc. Civ. Engrs*, SM1, 73–110.

Penman, A. D. M. & Watson, G. H. (1967). Foundations for storage tanks on reclaimed land at Teesmouth. *Proc. Instn Civ. Engrs* 37, 19–42.

Thorburn, S. & MacVicar, R. S. L. (1968). Soil stabilization employing surface and depth vibrators. *The structural engineer* 46, No. 10, 309–316.

Thorburn, S. & MacVicar, R. S. L. (1974). The performances of buildings founded on river alluvium. *Proc. Cambridge Conf. on Settlement of Structures*, Paper V/12.

Watson, G. H. & Thorburn, S. (1966). Building construction on soft alluvium. *Civ. Engng and Public Works Review* 61, No. 716, 295–298.

Discussion

Technical Editor: A. D. M. PENMAN*

DYNAMIC COMPACTION

L. Ménard (Techniques Louis Ménard)

The term 'dynamic consolidation' has been used for five years, but originally the term used was 'compaction'. The first job was at Ollioules in the South of France. We were construction engineers with the difficult problem of building five- to seven-storey buildings on a new fill of silt and sand, 5–8 m thick, reclaimed from the sea. The problem was the settlement of the whole platform even if deep piles were used under the buildings themselves, so the idea was to stabilize the whole platform. To do this we thought first of static consolidation. We tried an area surcharged to 10 t/m² and the settlements were 20 cm. We thought of dynamic consolidation by using a hammer of 8 t with a 10 m drop. The immediate settlement was about 50 cm. Later on we used a 15 t hammer with a drop of 20 m, with an immediate settlement of 90 cm. The buildings constructed on the site have a bearing capacity of 3 kg/cm² and settlements have been measured at 30 points. Even to date the settlements amount only to a few centimetres.

To begin with, we were working with pervious soils: with pervious soils there were no surprises. There were no surprises with the influence of depth: if we had to increase the depth by two we had to increase the energy of the blow by four. If a normal 20 t hammer dropped 20 m is necessary and acceptable to work a layer 15–18 m thick, we should use a 40 t hammer with 40–45 m drop to achieve densification at a depth of 30–40 m. Later this year we will use a 200 t hammer to consolidate 50–60 m of clayey silt. The anticipated immediate settlement is in the range of 2 m. There was only one surprise, that the influence of the speed is more important than the weight itself. When the weight is static on the ground, its effect spreads into the ground at 30–45°, but when the rammer speed is increased the energy enters the soil almost vertically. For this reason we have had to develop new types of free-fall equipment for several hundred tons.

The second thing is the problem of impervious soil. Four years ago I hoped we would not have to try to consolidate impervious soils: no-one would be stupid enough to ask us to do such things. We have had many surprises. I had not thought that saturated impervious soils could be consolidated in only a few minutes. But soils are not saturated, even 30 m below the sea; they contain 1–4% by volume of gas. It is difficult to measure this gas, but the immediate settlements of 50–60 cm indicate its presence and it can be seen coming from the surface. Low permeabilities of 10^{-6}–10^{-8} cm/s relate to static conditions, but become meaningless for dynamic conditions, especially when applied to boiling silts or boiling clays; it means that the pore-pressure is very close to liquefaction pressure. In this case we cannot speak of permeability to water of a solid structure—it is more a case of particles in water.

The third surprise is thixotropy of the soil. When there is no change in water content or density of the soil it can become stronger with time due to thixotropy.

Our aim has been to try to understand and improve soft soils. Some things that should work one way according to theory actually work in other ways and this applies even to instrumentation. We have been greatly confused by instrumentation. Piezometers have been

---

* Building Research Establishment.

Many of the contributions to this discussion were made at a British Geotechnical Society meeting held on 29 May, 1975.

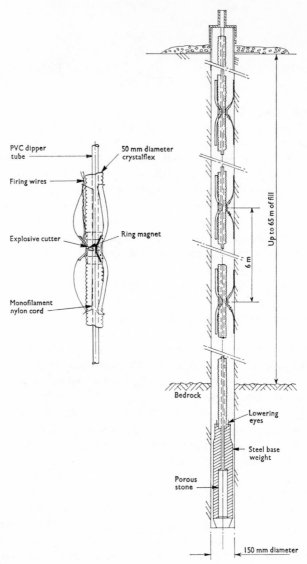

Fig. 1.   Borehole settlement measuring system

working correctly, but apparatus to measure settlements at different depths has given mislead-ing results.   The plates have moved up and down depending on the relative density of the equipment.   Even measurement of pore-pressure due to the shock has inertia problem which causes voids between the instrument and the soil.   There are no problems in sand and gravel, but several with silt, and we have to improve the techniques of instrumentation and measure-ment to weigh the results against observed performance.

J. A. Charles (*Building Research Station*)
    Research into the performance of buildings founded on land reclaimed from opencast mine workings is being carried out by the Building Research Station.   In Northamptonshire large

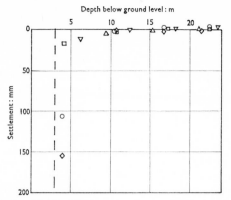

Fig. 2. Dynamic consolidation of a cohesive fill

areas of land around Corby have been worked for ironstone by opencast mining and in association with Corby Development Corporation we are investigating methods of treating these restored opencast sites so that conventional two-storey housing can be built on them.

In the area presently under investigation the ironstone was overlain by oolitic limestone and boulder clay. Excavation by dragline was so carried out that, in general, materials were replaced in their original sequence, i.e. with the clay above the limestone. At this site the fill is some 24 m deep.

To monitor the effectiveness of different methods of ground treatment borehole settlement gauges were installed in the areas to be treated. The magnet extensometer system used was basically that described by Marsland and Quarterman (1974) (Fig. 1). The settlement gauges were installed in 150 mm diameter boreholes drilled through the 24 m of fill into bedrock. Circular magnets which act as markers are anchored to the sides of the boreholes by strong springs. They can subsequently be detected by a reed switch sensor lowered down a central rigid access tube which is isolated from the ground by surrounding it by a flexible helically reinforced outer tube. During installation, when the apparatus is being lowered down the borehole by a rope attached to the heavy base-weight, the springs are held back by a monofilament nylon cord which can be cut by a small explosive device when the spring is in the correct position. The base-weight contains a coarse porous stone so that the central access tube can be used to measure the level of the water table as well as for settlement measurements. After installation the space between the outer flexible tube and the borehole walls was backfilled with dry sand. Settlements are measured by lowering a reed switch down the access tube with a steel tape attached.

An area of the restored ground about 50 m × 50 m has been treated by dynamic consolidation. The contractors, Cementation, applied the compaction using a 15 t weight falling 20 m. Initially a 10 m grid was used and at each point on the grid repeated blows were used to produce holes about $2\frac{1}{2}$ m deep. These were backfilled with the surrounding material. A number of similar stages of tamping followed repeating the process on the same grid and then using a new grid offset by 5 m. Finally there was a general tamping of the whole area with a reduced fall of the weight.

This method of tamping must result in the surface layers of fill being completely rearranged to a depth of at least $2\frac{1}{2}$ m, probably deeper. The settlement gauges were not directly tamped on and consequently showed heave in this top $2\frac{1}{2}$ m. The measured heave was as much as 300 mm.

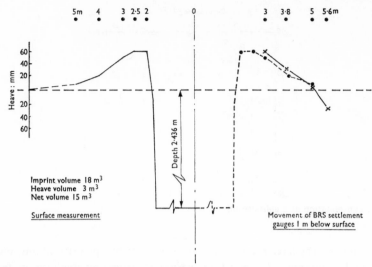

Fig. 3.   Corby Snatchill: imprint, 15 000 kg dropped 20 m 13 times

The settlements measured below 3 m on the different gauges are shown in Fig. 2.   A considerable variation in settlement was measured at 4 m depth; the maximum settlement of 15 cm was measured at a gauge only 1 m away from a primary tamping point on the 10 m grid and 11 cm of this settlement occurred during that first stage of tamping.   The average settlement produced at 4 m depth is 9 cm and below 10 m depth all settlements are smaller than 0·5 cm In the six months following dynamic consolidation further movements of about 0·5 cm have been monitored.

It is constructive to compare these settlements produced by dynamic consolidation with those resulting from preloading by a 9 m surcharge of fill on an adjacent area of the same site. The preloading experiment has not yet been completed, but measurements so far indicate an average settlement at 4 m depth in the restored ground of about 25 cm and 3 m settlement at 10 m depth.

R. W. Pearce (Cementation Ground Engineering Limited)

I should like to supplement the information presented by Dr Charles with data obtained by Cementation during the same contract.   Dynamic consolidation was carried out using a 15 t tamping block with a base area of 4 m². Initially impacts were given on a 10 m grid from a drop height of 20 m followed later by a pattern of overlapping imprints from a smaller drop height.   With the exception of the actual BRS gauge positions the whole site surface was therefore tamped.

There is good correlation between the heave measured on the BRS gauges and that measured at the surface from these pins.   Fig. 3 shows a section through the imprint which was closest to BRS gauge No. 4 some 3 m away.   On the left hand side the heave as measured from steel pins driven 0·5 m into the ground is shown, measurements being taken with an ordinary dumpy level.   The surface heave was measured north, south, east and west of the imprint but only the average is shown here.   The shape and order of magnitude of heave compares very favourably with the BRS settlement gauge measurements at 1 m depth below the surface. BRS gauges shown at 3·8, 5 and 5·6 m are not close to the particular imprint but were at these distances from the centre of other imprints on the site.   Although the measurements are not precise it is interesting to note that the order of heave volume to gross imprint volume is only

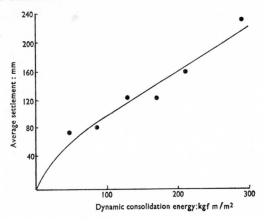

Fig. 4. Corby Snatchill: induced surface settlement

17%. The first pass imprints were on a 10 m grid and the net volume change would indicate an average surface settlement of 150 mm. In fact most of the imprints were about 2 m deep and the average first pass surface settlement was measured as 80 mm. Fig. 4 shows measured average surface settlement plotted against the input of dynamic consolidation energy. The treatment was designed to ensure that a compacted 10 m deep raft was formed in the top layers of fill sufficient to carry the envisaged housing loads. Although a final surface settlement of 240 mm was measured further energy would have resulted in further induced settlement.

Not all this induced settlement of course relates to compaction effect and an inclinometer installed 3 m from the boundary of treatment indicated lateral movements of up to 40 mm occurring at depths of 8 to 12 m. Other inclinometers further away showed no significant movements. It has been suggested that the lateral movement resulted from a hard layer in the fill below the inclinometer depth but site data is not conclusive on this point.

Only BRS gauge No. 5 was close enough, 5·6 m away, to a pressuremeter test position to make any comparison between increase in ground strength and induced settlement. Fig. 5 serves merely to indicate the order of improvement. Unfortunately settlement by BRS gauge was measured at only one point within the improvement zone. The post testing was carried out six days after the final pass treatment and even at this early stage the pressuremeter results show a factor of improvement of between 1·5 to 2.

The Corby data describe the compaction of dry clay fill and I should like to make some observations on the treatment of saturated silty sands by the process. Fig. 6 gives the average results of pressuremeter tests carried out under one of the tanks at Teesside where dynamic consolidation was used to improve the ground characteristics. The improvement achieved above the water table does not differ significantly from that achieved below it, the factor of improvement being approximately 2 in both cases. These results were typical of another three tanks that were in this area of the site. The induced settlement associated with this improvement as measured at the surface was 260 mm.

K. N. Engel (Elmat-Shand Consortium for Seismic Compaction)

Like all pioneering work, that of Ménard in dynamic consolidation has laid the foundation on which others might build, improve and add sophistication.

I should like to mention some of the improvements, which may be put into two categories: development of pieces of equipment which helped to improve the efficiency of operations and,

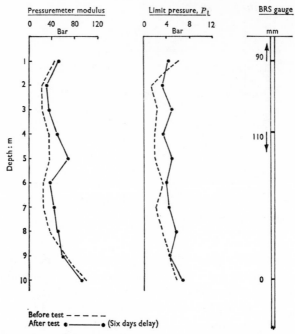

Fig. 5.   Corby Snatchill: pressuremeter results, BH B9

more importantly, a different recognition of the problems, which led to different control testing methods and procedures.   From experience in site investigation, it is apparent that whereas a certain number of boreholes per acre would give a tolerably reliable picture of soil conditions where the soils had been naturally deposited by geological agencies, a multiple of the number of boreholes was required when dealing with fills, and even then there would always be the doubt that some significant variations had been missed.   In heavy tamping one usually dealt with fills of one kind or another, whose characteristics were to be changed in a controlled way; thus it was vital that all significant variations in the material be investigated.   Consequently, to be certain of results, it was not sufficient to multiply the number of discreet points of investigation, be they borehole, pressiometer, vane, or penetrometer positions.   One had to employ some method of testing in the mass, so as to scan the whole mass of the ground, to map all zones of different characteristics and thus carry out detailed tests at significant locations.

A seismic surveying method could fulfil these requirements and Dr Dash of Imperial College, London, was approached.   His results of work on a similar problem were indeed applicable and he was able to work out a special survey method—the computer program protected by copyright—which fulfilled all requirements.

Elmat-Shand therefore had a very integrated method of carrying out and controlling this work to which the name Seismic Compaction was given.   (The word 'Seismic' referred primarily to the fact that it was seismic energy which caused the compaction).

B. P. Dash (*Department of Geophysics, Imperial College, London*)
On sites where compaction or consolidation of ground or fill is required, testing of the results is necessary to ensure the safety of any structures sited on it.   The engineering properties needed to be known, particularly strength and relaxation, can be measured by a variety of mechanical and electrical methods, of which borehole drilling, pressiometer and penetrometer

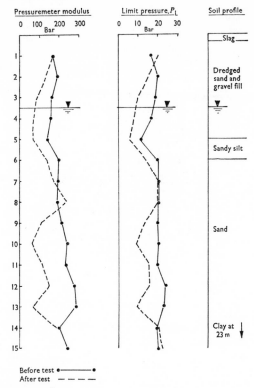

Fig. 6.   Teesside: tank 1, average results

testing are most used: but all are based on observations at discrete points, and extrapolation of the results is necessary to determine the condition of the site as a whole.   Properly sited boreholes provide information from which a picture can be derived that is acceptable in general, but which is particularly unsatisfactory where large areas are concerned.   In fills, it is quite inadequate because of the problem of lateral variation, but with no better methods available it has been widely used and its drawbacks compensated for by an increase in the safety factors used in calculations.   Improvements in reliability and coverage of such survey results have obvious cost advantages.

The seismic method described in this contribution provides a continuous survey of the acoustic properties of the ground, not restricted to discrete points of observation, from which a complete picture of the consolidation characteristics can be derived before and after compaction.   Additionally, as a routine procedure it predicts the effects of surface waves or 'ground roll' on any nearby structures during the compaction process.

The principles involved in the new seismic technique are almost the same as those of conventional shallow refraction seismic work undertaken for foundation work, well known to civil engineers, but the method used is varied in three respects.   The acoustic source used is the impact of the weight employed in the compaction process itself, instead of the conventional hammer device or dynamic charge.   Three-component geophones, which detect movement in vertical, horizontal-radial and horizontal-transverse directions, are used in conjunction with an array of 12 equispaced in-line geophones.   The spacing of the geophones is dependent on the depth of consolidation to be investigated.   The first group of geophones provides data for primary (P) and shear (S) wave velocities, energy and particle motion of the surface waves not

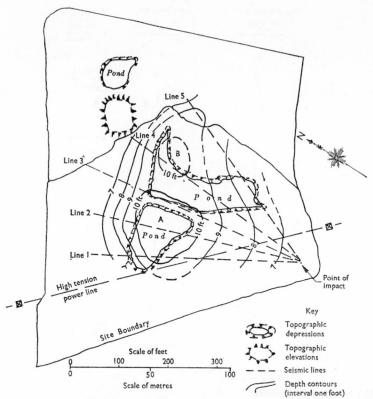

Fig. 7. Site plan and survey layout, near Basingstoke, Hampshire, showing depth contours in feet below road datum of 170·5 ft above msl at the point of impact. A and B are depressions in the interface

only at the point of impact but also at a predetermined point away from it. The P and S wave arrival times at individual in-line geophones, converted into real times and depths, are used to construct depth contour maps of structural boundaries between different acoustic velocities, before and after compaction. At each geophone location, therefore, detailed information is obtained about depths comparable to that provided by a borehole. The numerical analysis of the data involves the use of signal and noise statistics with correlation theory (Dash and Obaidullah, 1970) and the autocorrelation matrix method (Dash and Haines, 1974).

The improved seismic technique was put to practical use in a survey carried out on a site near Basingstoke, Hampshire, where ground consolidation was being undertaken by the Elmat-Shand Consortium for Seismic Compaction. The site, about 2·2 km² in area, lay in a topographic depression and became waterlogged during wet periods; when dry, it displayed a series of small ponds. It was located near the southern margin of the Reading Beds (Eocene) outcrop, with chalk occurring at the surface about one mile away and underlying the site itself at a depth of about 21 m. The Reading Beds are here characterized by poorly consolidated laminated clays, sands and mixed pebble and shell layers.

Fill had been placed on the site before consolidation, and consisted of similar material freshly scraped from an adjacent area which was also intended to be developed as part of the estate.

A seismic survey was carried out over the site using the new method just described, and repeated after the consolidation process had been done. Fig. 7 shows a depth contour map on the base of the poorly consolidated material before consolidation; zones A and B are

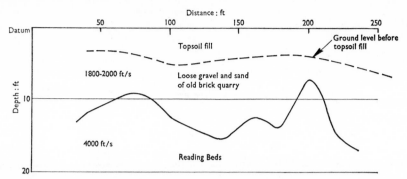

Fig. 8.   Depth contour along line 2 (Fig. 7) before consolidation, showing the acoustic velocities of the subsurface and the Reading Beds

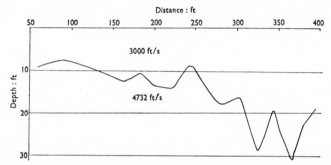

Fig. 9.   Depth contour along line 2 (Fig. 7) after consolidation.   Note the changes in the velocities and depths The inverse peaks indicate the zone where two or more 'thumpings' have overlapped

depressions in the base and it was noted that these did not correspond to the local topography, in particular the dry-season ponds.   As a result of this observation, further consolidation work was recommended over those areas where the acoustic velocities had been changed from 1700–1800 ft/s to about 3000 ft/s by the consolidation process.   The depth to the Reading Beds boundary and its velocity were also slightly altered by this process (Figs 8 and 9).

The boundaries between layers of different acoustic velocities are shown in profile in Fig. 8, and profiles over the poorly consolidated zones A and B are given in Figs 9–11.   These show the disappearance of the very low velocity layer after consolidation, giving rise to a thicker layer of higher velocity by combination with the adjacent layer.   The inverse peaks shown in the profiles represent locations where the effect of the impact of the consolidation weight has over-lapped between two impact points, resulting in deeper consolidation.

The decay of the pressure pulse with time, which is related to the distance from the point of impact, the acoustic source, is shown in Fig. 12.   Fig. 13 gives the relationship between the group and phase velocity with respect to frequency; this is important for the prediction of the effect of surface waves generated by the impacts on nearby buildings.

The significant advantages of this method are summarized in the following paragraphs.

Estimated costs for the use of this method are a fraction of the costs involved in a conventional drilling and sampling operation.

Data obtained from a borehole operation come from discrete points on the site without, in many cases, any realistic interpolation being possible.   The seismic method investigates the whole area systematically, each geophone giving data to produce a depth profile and thus giving

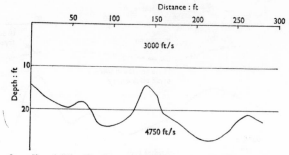

Fig. 10. Depth contour along line 4 (Fig. 7) after consolidation.   Note the acoustic velocities similar to those of Fig. 9.

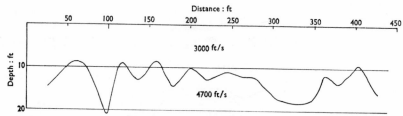

Fig. 11. Depth contour along line 5 after consolidation.   Note the acoustic velocities similar to those of Fig. 9

many more data than could have been produced by drilling at the usual spacing, together with data on the ground between the geophone points.

The seismic technique is used to give a before-and-after picture of the consolidation conditions of the ground, utilizing the weight that is actually used to produce the consolidation.

The method predicts the effect of the 'thumping' on any adjacent buildings, thus minimizing the risk of damage and financial claims.

The technique uses portable instruments, is easy to operate and can be repeated where necessary; no explosives are used.   In terms of cost, expenditure is minimal when compared with that needed to obtain the same information by borehole testing.

Acknowledgement.   I wish to thank the directors of Elmat Geophysical Ltd and Elmat-Shand Consortium of Seismic Compaction for their permission to publish the data obtained in Basingstoke.

N. B. Hobbs (*Soil Mechanics Ltd.*)

Dynamic compaction, or consolidation, is not as recent an innovation as many people have been led to believe.   The method was used in Durban, South Africa in 1955 to compact a loose hydraulic fill in order to support a 250 ft diameter crude oil tank.

Dutch soundings had shown that there were two zones within a depth of 40 ft where the slightly silty uniform fine sand was critically loose (the rods had penetrated virtually under their own weight).   The water table was 5 ft below the surface. An estimate of the costs of vibro-compaction and sand piling using open-ended driven casings had shown that a sum well in excess of £10 000 would have been required for peripheral treatment alone.   Consequently it was decided before embarking on any great expenditure to try to improve the ground by shock treatment using a 6 ft concrete cube, borrowed from the breakwater, and a walking crane.   The block was dropped 12 to 15 ft each time and the whole area of the tank covered

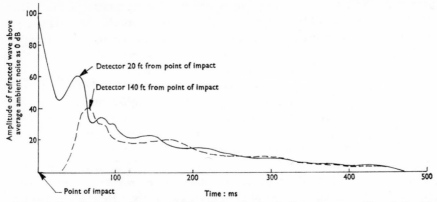

Fig. 12. **Decay of the pressure pulse calculated from the amplitude of refracted waves. The relationship between time and distance is calculated with an average velocity of 3500 ft/s**

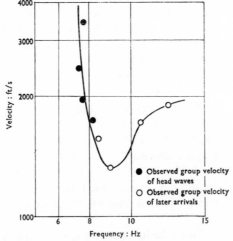

Fig. 13. **Relationship of group velocity of headwaves and later arrivals with frequency in Hz. The later arrivals have higher frequencies and as such do not contribute to the ground roll**

twice in about two weeks.   The maximum settlement was 13 in. with comparatively little surface loosening.   Further penetration tests made at the edge of the area showed that the effect had penetrated to a depth of 40 ft and that the critical zones had disappeared.   The cost was about £400.   No further improvement was necessary and the tank settled something over one inch during the water test.

A. W. Bishop (Imperial College, London)
    I understand that gas in soil is an essential ingredient of one of the hypotheses considered here.   Several years ago Dr Penman joined me on my yacht and we examined the behaviour of the mud forming the bed of the River Swale.   At that time I was interested in changes in pore-pressure due to rapid drawdown, and as the cheapest way of getting a rapid drawdown twice a day is by making use of the tide, we put our piezometers below the bed of a tidal part of the river (Bishop, 1966).
    What interested and surprised us was the amount of gas which came out of the river bed at low tide.   When the weight of the water was removed we could hear the gas bubbling up on

the bottom of the boat that we were using as our floating piezometer house. However, it apparently made surprisingly little difference to the pore-pressure changes. The response to a change of water load was almost immediate and gave a $\bar{B}$ value of about 0·96, although there was a good deal of gas there which could be seen and heard, and detected in de-airing the piezometers. I think such soil will behave from the mechanical point of view largely as though it were a saturated material, i.e. no change in effective stress due to a change in total stress. I don't know what that will do to the hypothesis referred to earlier, but it is a regrettable fact that in soft clays you can have quite a departure from full saturation—quite a lot of gas in them —but they will behave as if they were virtually saturated from the point of view of generation of pore-pressure and the volume change associated with the gas is probably remarkably small.

The second point I wanted to raise was one arising out of curiosity, not having seen this technique in operation. I recall a paper in the *Canadian Geotechnical Journal* some years ago by Meehan (1967), on the compaction of ground by elephants. He made a serious scientific study of this apparently promising technique and found that elephants were singularly ineffective because they behaved as socially well adjusted animals (once the one female elephant was removed from the team) and each trod in the footprint of its predecessor as it tracked across the site, and thus proper coverage could not be achieved. I wondered, if you wanted to put dwelling houses on top of the site, what would be the point of dropping a large weight at 2 m centres, if the building were to be only 10 m across. Surely one wants to get better coverage than this in relation to small structures. Would it not be better to build up a compacted crust with, perhaps, a lower energy and at much closer centres before putting on a larger impact to get deep penetration?

A third point is that at the site of Corby which appears to be partly saturated and where there may be more scope for compaction by decreasing the voids on an instantaneous basis, there may be the danger that if someone inundated the site at a later stage by altering the groundwater level conditions, they might cause much more important changes in volume than had been caused by the compaction. This initial settlement might thus give a misleading impression of site improvement, if one did not, at the same time flood the material, as one does with rockfill, to allow the natural decrease in void ratio which occurs due to the reduction in strength of the lumps.

It would also be of interest to know the influence on horizontal displacements, particularly at depth, when such apparently isolated drops are used.

J. D. Green (*J. Kenneth Anderson and Associates*)

Would Mr Ménard tell us how close his equipment has been used to existing buildings, and what precautions were necessary?

L. Ménard (*Techniques Louis Ménard*)

I shall try to answer some of the questions. How close to a building can compaction be carried out? There are two aspects: technical and social. From the technical point of view, considering general information, we can work 30 m from a building, but if it is not a sensitive building, we can approach to within 10 m. We have worked 5 m from a bridge on piles, but in this case we had to make a trench to protect the bridge from Raleigh waves. We have worked inside bulkheads, we have worked 1 m from a high block. So there is a wide range. Sometimes we have had problems at 100 m because there was a computer which could not tolerate even 1 or 2 μm of tilt. I would say that the major disadvantage of dynamic consolidation is that in some cases we cannot work close to buildings.

Regarding the social aspect, people do not like to have a big machine working very close to the window.   To have the sight and noise at the same time, is much worse than to have only the noise.

In reply to Professor Bishop's question on why we work on localized points, at first we worked as uniformly as possible, i.e. one blow per square metre, but found penetration of energy far better when blows were concentrated on points.   It is essential to compact the deep soil first and the surface soil second.   It would be bad to have only the deep compaction from points, but worse to have only surface compaction with compressible soil below.

One difficulty of dynamic consolidation is to approach the technique from the point of view of the fill.   It is difficult for the engineer in charge of a big project, to specify a technique that is difficult to understand.   We know by experience that it takes years of work for an engineer of the firm to really be in a position to understand all aspects of dynamic consolidation.   In addition to the technical aspects, there are all the specialist tools needed at the beginning, during and after the job: these include the dynamic oedometer which tells us about the amount of energy to liquefy or not to liquefy the soil at various depths.   We also have to assess the time between applications of dynamic compaction to get best results: it could be one day or it could be one month.

VIBRO-FLOTATION COMPACTION IN NON-COHESIVE SOILS

D. A. Greenwood (Cementation Limited)

Vibroflotation originated in Germany in the early 1930s.   The inventors of the basic machine were the firm of Johann Keller of Renchen with whom Sergey Steuerman was associated.   The first Keller machine patent was published in 1933.

· Steuerman was thinking originally of a machine for compacting mass concrete for dams. At about the same period in Germany there was much interest in vibrating equipment for compacting loose sands for the autobahn programme.   This interest is expressed by Loos (1936) in a report on various compaction methods to the First International Soil Mechanics Conference.   Trials were made with surface plate vibrators working on ground saturated by pumping water through boreholes.   Loos mentions that large scale tests were then running 'for huge deposits in boreholes.   The compaction is very high but this method requires permeable soils'.

In 1935 foundations were provided for a large building in Nuremburg resting on 16 m of loose fine sand over rock.   Cased borings 0·4 m diameter were made at 1·5 m centres and the sand was packed in 0·5 m layers as the casings were withdrawn.   The bearing capacity was doubled but the process was uneconomic.

It occurred to Steuerman that a robust and powerful vibrator could bore itself into the ground.   Accordingly in 1937 Keller undertook the first true vibro-flotation in the style of today for a building in Berlin on 7·5 m depth of loose sand.   Again, bearing capacity was reported as doubled with relative densities increased from 45% to 80%.

About that time Steuerman went to the United States of America and thereafter the story of vibro-flotation splits into the separate developments pursued by Keller and Steuerman.

In Germany, Keller continued to develop and exploit the process through the early part of the war and there is reference by Scheidig to sand compaction for grain elevator foundations in 1940.   Through the 1950s this development continued in Germany and by 1960 Keller was undertaking sand compaction jobs in a number of overseas countries and had developed the ability to penetrate to considerable depths, for example 21 m at a Dunkirk steelworks.   Futhermore, the enterprising engineering management of Keller led to the application of vibroflotation for a variety of novel projects, for example installation of large cruciform concrete anchors through sands to prevent flotation of a dry dock in Emden; the installation of offshore dolphins on sand foundations and even erection of a small lighthouse sunk into the sand by attaching external vibroflots which on release also compacted the sand round the foundation in situ. All these developments required linkages of several machines, up to six being used simultaneously for the lighthouse project. Keller also has succeeded in forming diaphragms to restrict horizontal permeability of laminated alluvia by using the vibrator at very close centres to mix the layers. Cement injection has also been tried but was not very successful since there are better ways of doing this.

Meanwhile, in America, Steuerman established himself as a consultant and also found Vibro-flotation Foundation Company of Pittsburgh and was able to use the process in 1948 for sand compaction for construction projects and for trials at Enders dam by the US Bureau of Reclamation. Throughout the 1950s the efficiency and economy of the compaction process was increasingly accepted and it became widely used all over the United States, but notably in Florida. This continues today.

The Americans, like the Germans, also achieved considerable penetration depths, for example to 20 m at the Hampton Roads Tunnel in 1954 and 25 m at Port Mann Bridge in Vancouver. However, the Americans have tended generally, for reasons of economy, to adopt somewhat shallower depths of treatment than the Europeans, who appear to be more security minded with respect to settlements of sands.

Regarding the introduction of vibro-flotation to Britain, Steuerman and VFC made an arrangement with Taylor Woodrow in 1956 or thereabouts to employ the American system. However, by 1957 they had sold it to Cementation who developed the process for British conditions. One of the early changes was conversion of the American machine to hydraulic drive to increase power and field reliability—the original machines were electric, designed for the American 60 Hz electricity supply system and hence required a generator for British operation. This disadvantage did not apply to the Keller electric machine, which operates at 50 Hz.

However, the major stumbling block in Britain was the lack of truly cohesionless permeable soils which restricted the use of the process. This gave Cementation impetus to consider the economics and the possibilities of backfilling with heavy stone in finer and more cohesive soils.

Their first practical trial of the system was in fact adopted as a site expedient for dealing with some 2 m thickness of organic silts encountered unexpectedly on an otherwise sandy site for six storey flats in Lagos in 1960. Thereafter the technique was deliberately applied for tests using 5 m and 6 m square concrete plates loaded to 20 t/m² at Llanwern and Teesport. The latter was successful and led ultimately to a succession of contracts for oil tank foundations over a number of years at the Teesport Refinery. This application for oil tank foundations is a special case in that it has been applied successfully to very soft soils—notably in the Venetian Lagoon where alternatives had been either too costly or had failed.

The same problem of cohesive soils had been encountered by Keller, and again in Nuremburg about 1959 they attempted strengthening a silty clay by dumping boulders into a 2 m deep bore and ramming them into mutual contact with a vibroflot. Subsequently they tried grabbing out pits which were backfilled with gravel compacted in the vibro-flotation manner, as at Teesmouth in the UK in 1961, as reported by Penman (1967). However, this was relatively expensive compared with backfilling holes formed with the vibroflot directly and the latter system was soon adopted by both companies.

The Keller system was operated in the UK from 1962 by Caledonian Foundations and was quickly used in Glasgow for housing redevelopments. This dry technique was continued by GKN Foundations when they took over Caledonian. Cementation did not adopt this method until some three or four years later, preferring the security of the wet system.

However, their adventures extended into treatment of refuse, PFA, and brick rubbles (Greenwood, 1970), as well as in 1966 to an early attempt at columns as friction piles, as described by McKenna *et al.* in pages 51–59.

It was not until 1972 that the first stone column job in the USA was undertaken at Key West by Keller. Meanwhile, in Germany, the success of vibro-flotation had spawned a number of competitive companies, whilst elsewhere in the world, notably in Pakistan where vibro-flotation was adopted for a series of large barrage projects, the process has been used in almost all Western countries. Very little is known in the West of the Russian vibroflot except for its use for compacting hydraulic fill sand between the cofferdams of the Aswan High Dam. In Japan a form of vibro-flotation had been used beneath oil tanks at Niigata and withstood the 1964 earthquake, where adjacent tanks on untreated soil did not.

The late 1960s saw machine developments by VFC and Cementation independently, both producing more powerful machines than hitherto, whereas Keller devised means of introducing stone at the bottom of a borehole and increased downward thrust to improve their dry process.

Today there is a considerable volume of work worldwide. There are, however, several systems akin to vibro-flotation based on piling casing drivers using contra-rotating eccentrics mounted at the top of a casing in which an essentially vertical vibration is induced. Whilst these penetrate efficiently, they are relatively poor compactors. The essential feature of a vibroflot is the generation of lateral vibration near the buried tip with consequent compacting efficiency in cohesionless soils.

N. E. Simons (*University of Surrey*)

I should like to refer briefly to the successful use of vibro-flotation in the refounding of two large tanks at Fawley. The original piled foundations had failed, and the two large tanks were refounded on adjacent sites by floating the tanks to the new positions. The ground conditions at Fawley consist of 9 m soft organic silty clay, over about 3 m of gravel overlying the Barton Clay. The refounding consisted of removing by dredging the soft organic silty clay, replacing it with imported gravel fill which was compacted by vibro-flotation. The tanks were 79 m in diameter and 19·5 m high.

After a series of trials the spacing for the vibro-flotation was fixed at 2·6 m and the effects assessed by SPT, static Dutch cone tests and dynamic cone tests. On average, vibro-flotation increased the measured values by a factor of about 4. Detailed results are given by Bratchell *et al.*, 1975.

Table 1 shows the measured settlements resulting from filling the tanks with sea water. Central settlements were estimated as 63 and 57 mm for the two tanks—no instruments were placed to measure settlements at these positions—under a loading of approximately 215 kN/

Table 1. **Observed and deduced settlements of tanks 281 and 282 at Fawley**

	Settlement, mm		
	In Barton Clay	In fill	Total
Perimeter tank 281	76	32	108
Centre tank 281	184*	63*	247
Perimeter tank 282	57	32	89
Centre tank 282	95*	57*	152

* Deduced values.

m². Predicted settlements using different methods based on the SPT ranged from 7 mm to 33 mm. Dutch cone results, interpreted by the De Beer and Marten's method gave a predicted settlement of 110 mm and Schmertmann's method, 14 mm. Schmertmann's procedure, in general, has much to commend it, but in this particular case with the 79 mm diameter loaded area on only about 9 m depth of gravel, Schmertmann's influence factors can be subject to some error.

The settlement of about 60 mm was roughly twice that predicted by the standard Terzaghi and Peck procedure from SPT results which, in every other case that I know of, is conservative. Could this be due to the surrounding soft alluvium plus the underlying Barton Clay? To check this, finite element analyses were carried out assuming that all the foundation was gravel with values of E increasing from the surface to the base of the gravel from 15 to 300 MN/m² and a Poisson's ratio of 0·15. The effect on the predictions of the Barton Clay and the soft alluvium surrounding the cylinder of gravel was then investigated and it was found that the predicted settlements were increased only by about 5%. We can only say that, for some unknown reason, the observed settlements were greater than those estimated with the method given by Terzaghi and Peck and that this observation is not due to the presence of the soft surrounding material.

R. Sparks (*Cementation ground engineering limited*)

A moderate amount of work has been carried out in recent years to evaluate the effect of vibro-flotation techniques in various ground conditions. The published work on this subject generally deals with the effect of one particular type of machine, with its own individual characteristics. It is reasonable to suppose that the ground would react differently when undergoing treatment by different machines. I should like to present details of some work which was carried out to try and compare the effect of different machine characteristics on the vibro-compaction of cohesionless soils. Three different vibrating machines were used at a site in Ardersier, on the shores of the Moray Firth, Scotland. The site consisted of 3–4 m depth of hydraulically placed sand fill, overlying a generally dense sand, silty in places. Fig. 14 shows the envelope of the grading curves for the hydraulic fill. None of this material passed as BS 200 sieve. The sand was not particularly well graded, lying generally in the fine to medium range. It was considered suitable for treatment by the pure sand compaction technique of vibro-flotation, i.e. no imported stone backfill was used, the bores being filled by sand from the site surface.

A comparison of the basic vibrating machine characteristics is shown in Table 2. Essentially the vibroflot is the least powerful machine, whilst the powerflot is the most powerful of the three. The intermediate modified powerflot has a smaller eccentric in a powerflot body

Table 2. **Comparison of machine characteristics**

	Vibroflot	Modified powerflot	Powerflot
Outside diameter, mm	406	324	324
Frequency, rev/min	1300	1800	1800
Amplitude, mm	4·6	4	6·8
Out of balance force, kN	46	46	85

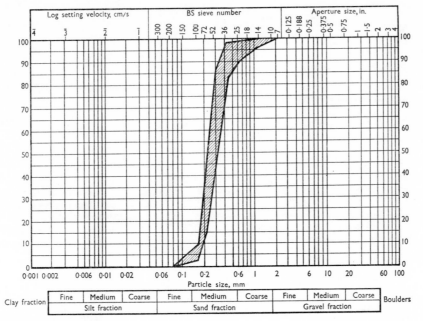

Fig. 14.   Envelope of grading curves

shell, resulting in a smaller amplitude of vibration and smaller out-of-balance forces.   Thus the energy imparted to the soil for each machine increases from left to right in Table 2.

One inch diameter dynamic cone penetrometer equipment was used to evaluate the effects of the three vibrating machines on the hydraulic sand fill.   Although this is not standard equipment in the UK, it was found to be a quick, reliable and economic method of comparing the effect of the different machines in reasonably homogeneous ground conditions.   Penetrometer tests were carried out before treatment by vibro-flotation and at various radii from the compaction centres after treatment.   Fig. 15 shows the penetrometer test results before treatment, and at various radii from the centres of compaction after treatment, using the vibroflot.   A considerable improvement in the degree of compaction can be observed up to radii well in excess of 2 m.   At a radius of approximately 3 m, the improvement would appear to be almost negligible.

The degree of compaction achieved at various radii from the centres of compaction by the three different vibrating machines is compared in Fig. 16.   Fig. 16(a) indicates that at a radius of approximately 1 m, the degree of compaction achieved by the less powerful vibroflot appears, somewhat surprisingly, to be greater than that achieved by the two more powerful powerflot machines.   At a radius of approximately 2 m (Fig. 16(b)), the degree of compaction achieved by all three machines was roughly similar.   However, at radii greater than approximately 2 m, the powerflot achieved a greater degree of compaction than the vibroflot (Fig. 16(c)).   It will be noted from Fig. 16 that the degree of compaction achieved by the powerflot was very consistent up to a radius of at least 2·7 m from the centres of compaction.

Fig. 17(a) suggests that compactions formed using the pure sand compaction technique with the vibroflot consisted of a relatively loose core of sand, building up to a maximum density at a radius of 1–1·5 m from the centre of compaction.   The density then rapidly reduced until at a radius of approximately 3 m, the degree of improvement was negligible.   Fig. 17(b)

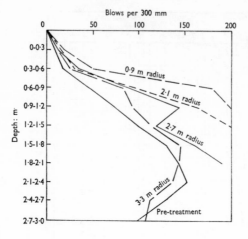

Fig. 15. Penetrometer test results at various radii from centres of compaction formed using vibroflot

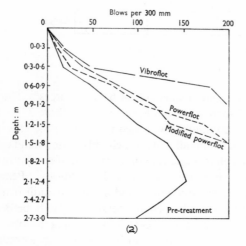

(a)

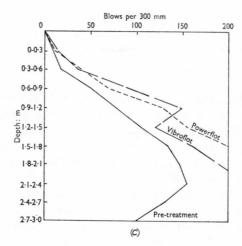

(c)

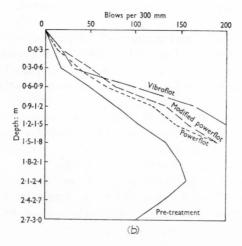

(b)

Fig. 16. Penetrometer test results at various radii from centres of compaction for different machine types: (a) post-treatment tests at radius of 0·9 m; (b) post-treatment test at radius 2·1 m; (c) post-treatment tests at radius of 2·7 m

illustrates that compactions formed using the powerflot had a denser core, with a more gradual increase in density to a maximum at a radius of approximately 1·5 m from the centre of
compaction.   Even at a radius of approximately 3 m there was still a considerable improvement over the pretreatment penetrometer test results.

The sectional density contours for the three different types of machine are shown in Fig.
18.   These diagrams again illustrate that the vibroflot tended to build up a cylinder of compact sand around a central relatively loose core, whilst the powerflot more closely achieved the
ideal uniformly stratified state across the profile section.   It should be emphasized that the
conclusions drawn in the foregoing represent the compaction effects of the three different
types of machine in one type of soil, using the pure sand compaction technique, without the

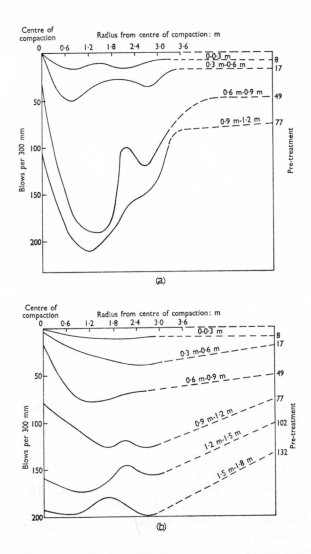

Fig. 17.   Penetrometer test results at various radii from centres of compaction: (a) vibroflot; (b) powerflot

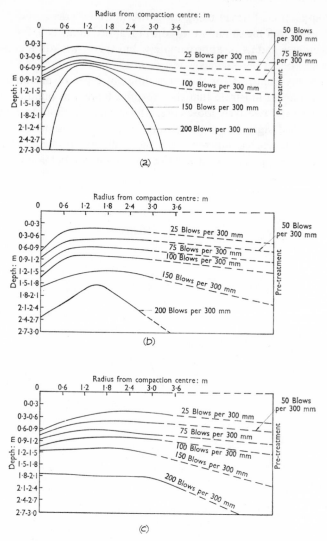

Fig. 18. **Penetrometer test contours: (a) vibroflot; (b) modified powerflot; (c) powerflot**

addition of imported stone backfill. The more sensitive silts and clays undoubtedly respond in different ways to the various machine characteristics, and this is a subject for further study.

E. H. Steger (Pell Frischmann and Partners)

I may be able to shed a little more light on the history of vibro-flotation. In the late 1950s my then employers put me to the task of Steuerman's patents. This meant reading up all relevant literature and publications going back to about 1934 and was quite fascinating. I was impressed, at that time, with the method which used a pipe outside the vibrator to inject cement grout whilst raising it and thus install an in situ column. It is surpising to learn today that this technique was not successful and had been discontinued. I witnessed a very impressive demonstration in Germany in 1958.

With reference to scientific design methods relating to dynamic compaction and vibro-flotation, these methods are generally used in totally heterogeneous soils and fill only, with engineering properties which are likely to change over a few feet in any or all directions. An attempt to derive scientific design methods for such crude construction expedients under the usual conditions would be misleading to the client and unfair to his professional adviser. The main thing is that the methods work under suitable ground conditions; the actual amount of compaction is likely to be achieved with an economic effort can only be established by a field test. I was sorry to notice the complete absence in the papers of any comments on, or measurements of, the damping effects of different soils for given impacts; this would seem to be a very important piece of knowledge as dynamic compaction may be carried out near to existing structures.

I should like to be enlightened on a point concerning the economics of vibro-flotation used for dynamic compaction. Ever since I used vibro-flotation in 1958 to improve the bearing capacity of some 60 ft^2 of Rhine terrace gravel over a very considerable area, I have been hoping to see the method in the UK. Its use here is somewhat restricted by geological factors —extensive sand and gravel deposits are not so common as elsewhere. I am not convinced of an equally successful application of dynamic compaction in cohesive soils quite apart from the groundwater problem usually associated with them.

However, my problem has always been the following: most sites for which the method might be used are parts of larger built-up regions with fixed levels for roads and services. If the ground surface in some areas is, therefore, lowered by 3 to 6 feet—this is the order of magnitude which one would expect from a successful dynamic compaction effort—the level has to be raised again by placing and compacting suitable fill. This can be very costly. Also and quite frequently, loose sands are overlain by alluvial clays which form a desiccated crust. For housing developments, this crust, which would of course be destroyed by compaction, can often be used as a reasonable founding stratum. I was most interested in Dr Simons' settlement measurements at Fawley. I suspect that most of the settlement took place in the Barton Clay below the gravel but should be grateful to have more details.

I. R. Clough (Cementation Ground Engineering Limited)

The absence of papers on vibro-flotation in granular materials does not indicate the demise of this process. In fact projects have been executed in all parts of the British Isles despite topographical conditions which have not been conducive to the formation of clean sandy soils appropriate to the process. However, overseas, the majority of the ground treated has been in granular materials demanding classic vibro-compaction, with recent emphasis on hydraulic fill schemes, particularly in the Middle and Far East. Frequently the need to guard against liquefaction of such granular materials in earthquake zones was a prime factor in the decisions to carry out compaction treatment. This generally involved raising the relative density of the in situ material to a minimum of 70% although it must be noted that every case must be evaluated on the assessed earthquake magnitude for the locality and local experience.

The trend towards higher powered machines has been noted in the discussions; these have been evolved to obtain greater radii of vibratory influence in granular materials and greater penetration rates in clays. The use of higher powered equipment has led in some cases to difficulties in dealing with material between these extremes of particle sizes particularly in the silt range where high power and amplitude can lead to destruction of the fabric of the soil which cannot then give support to the stone columns formed therein. This need not be a problem provided that power, amplitude and frequency of the vibrator used are matched with the soil to be strengthened and early site results are monitored closely to check that the correct choice has in fact been made, with prompt changes in equipment if this is not the case.

The accurate estimating of timing for works has been queried in this discussion. Potentially this can be difficult where fill materials are to be compacted especialy as in the UK some 70% of current compaction work is carried out in fills. These fills often vary in composition far more than naturally laid down material. The normal coverage of boreholes and the standard testing and sampling often does not reveal a representative picture in these circumstances particularly of fills comprising industrial and waste disposals. Inspection of a substantial number of trial pits (which can be done relatively cheaply with modern light excavation equipment) is an essential pre-requisite both for accurate estimation at the planning stage, and in the execution of the work. The success of this approach was shown in a recent analysis covering a two-and-one-half year period in which over 90% of contracts reviewed were completed within the programmed time irrespective of any delays from whatever cause.

M. P. Moseley (*GKN Foundations Limited*)

The principle of applying vibrations to cohesionless soils to achieve an increase in density and reduction in void ratio is well known. The resulting increase in bearing capacity or reduction in total and differential settlement can be of considerable value to engineers. The application of vibro-replacement and vibro-flotation techniques to achieve these ends in cohesionless soils has proved emimently successful in a large number of projects and countries. Indeed, it is worth remembering that the classical application of vibro-flotation was in loose sand deposits.

There are numerous examples in the literature of the successful application of vibro techniques to cohesionless soils, but I should like to show one recent example. My firm was contracted to undertake vibro-flotation work at Irvine in Scotland, where a refractory furnace and ancillary buildings were to be constructed. The site investigation had revealed a succession of medium dense sand, with N values in the range 15–25, overlying boulder clay at a depth of about 9·0 m below ground level. The N values exhibited variation both laterally and vertically over the site, and concern was expressed regarding differential settlement, particularly to the refractory furnace which was very sensitive to movement. A load test was performed on a 3·0 m × 3·0 m base, constructed on a treated area, the results of which are given in Fig. 19. A total settlement of about 4 mm was recorded at an applied pressure of about 320 kN/m². The proposed working pressure was 215 kN/m². I should note that the application of vibro-techniques in this instance was not so much one of increasing bearing capacity, as one of reducing settlement and thus providing an insurance for the foundations to the refractory furnace.

I think it is agreed that vibro-replacement and vibro-flotation produce an improvement in the engineering properties of cohesionless soils. In Berlin in 1937, relative density was improved from 32% to 80%. A paper on a 20-storey building in Lagos, Nigeria, records improvements in relative density from between 20–40 and 80%. At a sugar silo in Durban, South Africa, cone penetrometer point resistance readings were increased from about 500–1500 lb/in² to 2000–3000 lb/in² after treatment. At Lincoln, my firm recently undertook pressuremeter tests before and after treatment. Improvements in limit pressure from 5–10 bar to 10–20 bar were recorded, and increases in pressiometric modulus from 50 to 100 bar. There are numerous other instances of improvement in the literature but it would appear that,

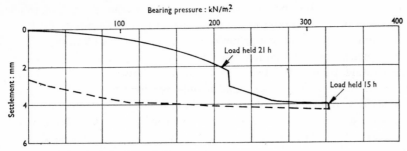

Fig. 19. Load test at Irvine on $3 \cdot 0 \times 3 \cdot 0$ m base

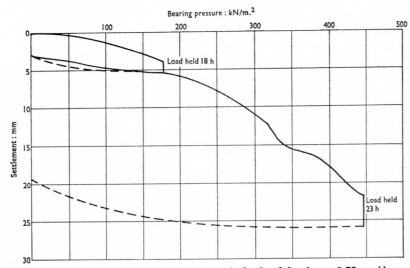

Fig. 20. Tower Hamlets, Area 2: full-scale load test on strip footing $3 \cdot 2$ m long $\times 0 \cdot 75$ m wide

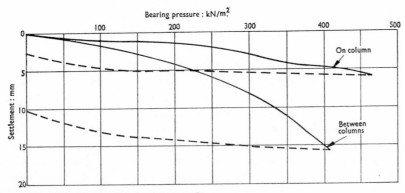

Fig. 21. Tower Hamlets, Area 2: plate load test results

whether the method of testing be by cone penetrometer, pressuremeter, plate tests, SPT or whatever, the order of improvement lies in the general range 2–5 times.

Finally, I should like to mention a group of soils referred to as fill in the UK. These soils present strongly heterogeneous characteristics and are difficult to classify, and are best tested by full-scale loading trials. My firm has recently undertaken work in Tower Hamlets, London for a housing redevelopment, and a full-scale load test result is given in Fig. 20. This test was performed on a strip footing 3·2 m long by 0·75 m wide and settlements of up to 26 mm were recorded at about 430 kN/m². The proposed working load was 165 kN/m². The soil conditions were about 4·5 m of loose fine black ash overlying sand and gravel. In addition, plate loading tests were performed on and between columns, and the results of these are given in Fig. 21.

The results presented from Irvine and Tower Hamlets both illustrate the successful improvement of granular soils. It must be emphasized, however, that a sound knowledge of the soils to be treated is of prime importance.

R. G. H. Boyes (*Andrews and Boyes Limited*)

It would be interesting to have comparative figures of the cost and performance of the dynamic consolidation, sand column and similar methods under discussion in comparison with the older and more traditional methods of forming a suitable tank settlement base. Such techniques include forming rush mats—a modern version of this employing synthetic fibres as rushes become harder to obtain—and ideas such as that of the Durley dome system. In the latter method, invented, I believe, several years ago in the USA by a Mr Durley, a pre-stressed or post-tensioned ring beam is built up first of all on the ground around the area on which the tank will rest. The middle area is then filled with sand/gravel and the idea is that the forces exerted by the tank will be taken up by the ring beam through outward pressure rather than downwards to cause settlement. Only one or two examples of the Durley dome appear to exist, one being at the Shellhaven refinery where it remained forgotten until fairly recently.

N. E. Simons (*University of Surrey*)

I have already mentioned the difficulty of trying to estimate the central settlements due to the fill itself of the tanks at Fawley. The observed settlement around the perimeter was about 32 mm in the fill alone, and the total measured perimeter settlement was of the order of 100 mm. At the centre of the tank a total settlement of the order of 200 mm was measured and my best estimate was that something of the order of 60 mm took place in the fill. It was certainly more than 32 mm, but equally it could have been less than 60 mm.

Mr Boyes asked about alternative forms of foundation and I can assure him that the consultants, Nachshen, Crofts and Leggatt, thoroughly investigated alternative types of foundation for these tanks. There was no question of founding the tanks on the soft alluvium, because they were 19·5 m high and the alluvium was very soft indeed.

It should be noted that at the time when a decision had to be made, there had been a failure of the piled foundation so that a foundation other than a piled foundation was to be preferred. The replacement technique that was adopted for the conditions at Fawley, with a favourable supply of fill material, was a rapid and secure way of founding the tanks which are now functioning satisfactorily.

D. A. Greenwood (Cementation Limited)

Referring to Dr Simons' remark that at Fawley measured settlements of the compacted granular backfill exceeded those predicted by the normally conservative Terzaghi and Peck method, I suggest the following possible explanations.

First, tank centre settlements on which the comparison was made were based on estimates from finite element models with E varying with depth but not laterally. I believe that a uniformly loaded flexible tank on a granular foundation may settle more at the edges than in the centre. Accordingly, whilst recognizing the benefits of finite element work, I query how precisely the prototype was simulated.

Second, cohesionless fill treated by vibro-flotation is in a normally consolidated condition after compaction. The Terzaghi and Peck rules, however, are based on empiric observations of buildings on natural soils, most of which are likely to be overconsolidated to some degree. Thus the rules may not be conservative for truly normally consolidated sands.

Third, at Fawley the fill covered a wide range of particle sizes up to 50 mm and the content of these coarse particles may well have increased the average level of SPT blowcounts analysed statistically.

Regarding effectiveness of compaction by vibro-flotation at shallow depths, whilst it is true for silica sands that with current machines the upper 0·5 m is not well treated, those working in areas of volcanic soils with low specific gravities should beware that the same machines may give a greater depth, say 1–2 m, of poorly compacted surface due to over excitation of light-weight soil particles.

J. Gray (City Architects Department, Manchester)

Ground stabilization using the dry process, was used by Manchester Corporation as early as 1963. The process was judged to be suitable for dealing with the problem of rubble filled cellars in slum clearance areas and in the event this early decision proved to be correct. Until the Keller vibrator became available all cellars were excavated and backfilled in compacted layers with limestone chatter and flyash—an expensive and slow process.

Using design information made available by Mr Thorburn, the Chief Engineer of Caledonian Foundations Ltd, and Mr Rappert, Managing Director of Johann Keller GmbH, Frankfurt, a system of compaction points was designed for each house type generally directly under load-bearing walls at 0·9 to 1·5 m centres. Because the slum clearance sites were in active urban areas of the city the dry vibro-replacement system was used throughout.

The design of the house foundations was based on a minimum safe bearing pressure of 155 kN/m² and plate loading tests were carried out regularly both directly on compaction points and in between them. Under test the ground was loaded to three times the working load, i.e. 4·50 kN/m² with a differential settlement between any two test points not to exceed 6 mm.

The substrata in the early slum clearance areas in Chorlton on Medlock were firm to hard marl so that any settlement registered during a plate loading test could be assigned directly to the top 2·5 m or so of ground or consolidated fill. Nominal steel reinforcement was provided in the strip footings as a safeguard but with increasing experience and confidence the reinforcement was gradually omitted. The first two-storey houses to be constructed on stabilized ground, usually in long blocks of six or eight dwellings have now been up for over twelve years and not a single report of brickwork cracks or other signs of undue settlement have been received.

When steel-framed schools with widely spaced columns were to be erected in the clearance areas of the city a different design technique was adopted. Again based on Mr Thorburn's figures a limiting load for an isolated stone compaction point of 100 kN was adopted to give a safety factor of at least $2\frac{1}{2}$ and stone columns were put in in small groups of three or four under each steel column according to the column load.

At a later stage GKN Foundations acquired the Keller vibrator and Cementation Ground Engineering Ltd, brought their powerflot into Manchester. With the help of GKN Foundations Ltd, a full-scale loading test was carried out on a group of compaction points formed in a typical Manchester clearance site, i.e. a basement area 2·4 m deep very loosely filled with brick rubble and bits of timber from the demolished houses with firm sub-strata below. The significant fact that emerged was that a well compacted stone column under these conditions was capable of carrying over 500 kN load before any appreciable deflexion took place.

As the clearance areas spread to the north side of the city a number of sites with soft clay and silt sub-strata had to be stabilized. It was found that a reduction in the spacing of the stone columns usually provided a satisfactory foundation but in very soft conditions the number of stone columns had to be doubled before adequate bearing values could be achieved. Recently this problem has been overcome more easily by reverting to the 'wet' process in areas where soft clays and silts have been found.

Two schools have been built on soft clay sites with values of undrained shear strength of about 30–35 kN/m². Groups of stone compaction points were placed under each structural column in a 1·22 m diamond pattern to ensure a safe ground bearing pressure of 150 kN/m² with a similar nominal pattern but more widely spaced (1·83 m) under the floors to achieve 100 kN/m². Both schools have been occupied for over six years and show no visible signs of differential settlement.

A similar system was used in the design of the foundations for a large technical college extension where the ground consisted of 2·5 to 3·0 m of variable fill and old cellars overlying medium hard clay with marl and sandstone at greater depths. Vibro-consolidation proved to be much cheaper than piling (the foundations of an adjacent elevated motorway at this point had bored piles taken down 20·4 m to the sandstone) and has proved effective in supporting isolated column loads of 500 to 800 kN on reinforced concrete pad bases.

As land within the city boundaries has become scarcer the possibility of building on old tip sites had had to be investigated. These old tips are usually abandoned clay pits which have been refilled indiscriminately over varying periods of years. Often they are 12 to 15 m deep. For two-storey housing construction piling is far too expensive and the experiment was tried of using the dry vibro-replacement process to a depth of about 3·66 m to provide a layer of hard material on which the houses could be built. So far the system appears to have worked satisfactorily but more investigation is needed. A dwelling for aged persons, two storeys high of load bearing brickwork construction has been built on 12 m of fill with apparent success by this method. After five years of occupation no signs of movement have been observed. As a result of experience a large housing estate is at present being constructed on a filled ravine where the fill varies in depth from nothing to 15 m. Again a layer of consolidated ground has been formed under each block of two-storey dwellings. Using measuring techniques developed by the BRE the foundations of a line of blocks of dwellings across the ravine have been fitted with standard BRE levelling points and two deep datum points have been established in the hard marl, at a depth of 9 m (30 feet) at either end. Measurements are being taken monthly with a Wild precision level and it is hoped to be able to show, in due course, the pattern of behaviour of this type of foundation.

Several 'wet' vibro-flotation jobs have been undertaken, particularly a housing estate on the Cheshire border where there were 300–450 mm thick bands of peat 1·5 m below ground level and a large development of five-storey flats in a clearance area with a very wet silty substrata.

Fig. 22

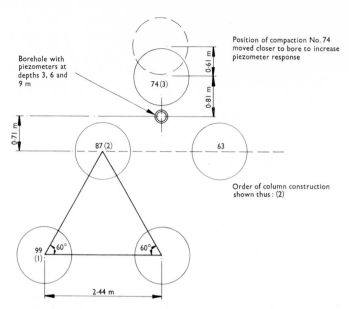

Fig. 23.   East Brent trial: piezometric observations during stone column construction (plan)

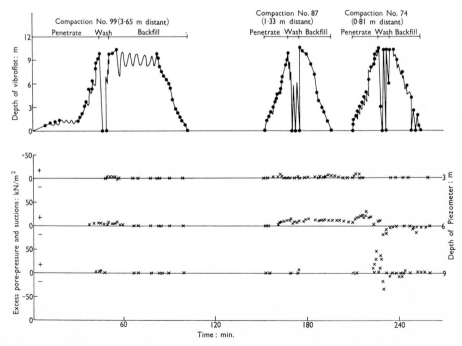

Fig. 24.   East Brent: piezometric observations during stone column construction

There may be technical arguments as to the validity of vibro-replacement or vibro-flotation but from a practical aspect it appears to work very well and is an economical solution. It is very rapid and reasonably cheap method of overcoming foundation problems provided the soil conditions are right. It enables difficult sites to be treated before the general contractor starts work and leaves him with a site on which he can build in traditional ways, i.e. strip footings at 750 mm below ground level. The general contractor is not delayed because the vibro teams always seem to be faster than the excavators and an efficient production of houses results.

Between 1963 and December 1974 over 26 000 houses, apart from schools, colleges and so on have been built in Manchester on foundations supported by vibro-consolidation and current contracts and tenders are in the region of £0·5 million.

STONE COLUMS IN SOFT COHESIVE SOILS

D. A. Greenwood (*Cementation Limited*)

The Paper by McKenna *et al.* shows how stone columns reduced neither rate nor amount of settlement and it suggests that this was due respectively to the pore space in the columns becoming choked with clay slurry during construction and to penetration of clay squeezed in when columns bulged during loading.

I do not dispute the facts reported in this paper but disagree with the postulated explanations. Concerning drainage through columns, I should expect the voids between normally uniform 38 mm aggregate to be filled during the construction process. However, with continuous water circulation whilst backfilling the column, water sorting of displaced clay occurs which removes much of the finer fraction, leaving predominantly silty sand in the void spaces (Fig. 22). Admittedly this has variable, and sometimes moderately low permeability, but by contrast with the soil fabric of a silty clay it is hardly of comparable low permeability.

Regarding blockage of soil drainage paths during column building, there is field evidence at East Brent of rapid pore-pressure dissipation after construction of columns. Fig. 23 shows the plan positions of columns Nos 99, 87 and 74 near the centre of the trial areal in relation to a borehole in which three piezometers were installed at depths 3 m, 9 m and 12 m prior to column construction and continuously observed throughout their construction. Columns were formed working towards the borehole at distances 3·65 m, 1·33 m and 0·81 m from it.

It is clear from the results observed (Fig. 24) that all piezometers were responding and that despite potential remoulding at the clay–column interface during construction, they all resumed equilibrium readings within minutes of restoration of equilibrium pressures in the column bore. Pressure fluctuations were generated by surging of the vibroflot rather than by shearing disturbance of the surrounding soil.

Regarding squeezing of clay into pore spaces of coarse backfill as columns dilate, there are three observations which suggest that this cannot have happened. First, penetrability of clay into pore spaces of the 38 mm stone can be calculated using relationships proposed by Raffle and Greenwood (1961). The method is based on the Kozeny–Carman equations relating permeability to an equivalent average pore radius for which shearing resistance to penetration can be calculated. Making very conservative assumptions of permeability $k = 10^3$ cm/s for the stone and a void ratio of 0·6 for dilated columns with unit coefficient of lateral earth pressure, the maximum penetration of clay into the column under influence of embankment loading is only 1·5 cm for the stated average cohesion of 26 kN/m². Reference to the original report by Soil Mechanics Ltd shows the average remoulded strength from in situ vanes was 7·4 kN/m² with a minimum in any one metre depth of 4·7 kN/m². The soil surrounding the whole length of the columns would have to reduce uniformly to this minimum strength to approach the necessary penetration to account for the volume of settlement saving—say 25% to 50%—expected from the columns on the basis of empiric observations elsewhere (Greenwood, 1970).   Since bulging begins and continues preferentially at a level where support of the column is weakest, penetration of the clay into the bulged length only would have had to be of order 20 cm or more.

The second point is more significant.   The lateral stresses due to embankment loading were in the range 180 to 240 kN/m^2.   Much higher lateral stresses are imposed locally on the stone by the vibroflot during column construction.   Any remoulding and penetration of the soil by the stone (leaving the same relative inter-penetration) would occur at this stage.   Subsequent external stresses could not then induce further penetration of soil into the column.

Observations of excavated columns through soft silt and clays show that voids between uniform stone particles are filled with sands and silts derived from the natural soil as described above.   The magnification in Fig. 22 of the edge of a column in silty clay shows sand filled voids and no significant penetration of clay.   In any event inhibition of drainage due to silting during construction is incompatible with squeezing in of clay under post construction loading. Local remoulding of sensitive clay might well occur during vibration but the thicker the 'pad' of stone between the soil and the vibroflot, the lesser is the excitation of the soil.   The presence of a skin of remoulded soil on the perimeter of the columns would not affect lateral restraint to a bulging column significantly.

I believe, therefore, that the inhibited drainage and squeezing clay hypotheses are nonstarters.   An alternative explanation must be sought.   Analysis of the critical length of column—the shortest length at which both bulging and end bearing failures occur simultaneously (Hughes and Withers, 1973)—for average $c_u = 26$ kN/m^2 shows this to be about 10 m, which is approximately the same as the column length.   The maximum load which the column can sustain without bulging is just sufficient to sustain 7·9 m of embankment, allowing for the minimum undisturbed soil strength—say 20 kN/m^2.   Thus the columns were only just of sufficient length to avoid end bearing failure as friction piles.   Using fully remoulded strengths the critical length becomes 32 m.   This shows that only slight remoulding along the perimeter of the column would be sufficient to allow a punching failure of the column acting as a stiff pile.   Maximum end bearing resistance was only about 1/7 of column loading.

It is important to appreciate that there was no clear boundary between the silty clays and the underlying 'sands' as shown in Fig. 2 on p. 52.   In three boreholes in the trial area Soil Mechanics Ltd was able to extract open drive samples at 1·5 m to 3 m intervals yielding moisture contents as high as 40–70% at depths below 12·5 m.   They contain a high proportion of silty clay and peat with sand predominating only near the bottom.

The columns were thus analogous to friction piles with the additional requirement that they should not bulge.   Support for this hypothesis is in Fig. 9 on p. 56.   Maximum excess porepressures in untreated zones were generated at depths from 2 m to 5 m embracing the major peat deposit.   This contrasts with the pressure profile in the stone column area which increases steadily with depth to the toes of the columns.   Had the columns dilated, excessively high pressures would have been expected first where lateral restraint was least (about 2·5 m depth in this case) resulting from shedding of stress to the clay.   Alternatively if the columns were relatively stiff and performed as piles with insufficient end bearing the relative shearing movement on the sides of the column would be progressively greater with depth and might well generate the observed profile pressures.   Warping of clay laminations would occur during this relative movement according to its local severity and might contribute to closing off drainage to the column, increasingly with depth.   It is worth considering whether warping could be more damaging to drainage than packing of stone against the borehole walls.   Finally,

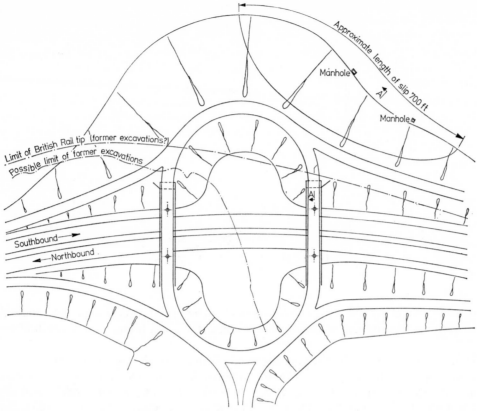

Fig. 25

it is clear that the columns performed properly as comparatively stiff reinforcing members. However, the spacing and depth adopted for the trial in 1966 were inadequate for the imposed loading and with uncertain knowledge of column behaviour then available this was overlooked. Current designs for stone columns as friction piles under widespread loads should account for the possibility of loss in strength of sensitive soils or otherwise provide sufficient end bearing. The fact remains that the treated area was not involved in the slip and the pore-pressure distributions clearly show that the columns enhanced stability substantially by forcing potentially dangerous pore-pressures (and hence weaker sliding surfaces) to lower levels.

G. H. Thompson (Cementation Ground Engineering Limited)

I should like to present a case history of stone columns used to stabilize a motorway slip road embankment, in a manner rather similar to that indicated in Fig. 3, p. 48, in the paper by Rathgeb and Kutzner, to stimulate discussion and further thought on the mechanism by which the stone columns achieve the increased stability.

The project concerned was the Hendon Urban Motorway and specifically the south-east slip road embankment at the Scratchwood Interchange near the southern end of the M1, constructed in 1967. Those who know the Scratchwood service area will recall its unusual plan of 'U' rather than 'O' shaped access roads. Fig. 25 shows the projected interchange, with slip roads leading to high level bridges crossing the motorway. The slip road pavement level at the eastern end of the bridges is some 50 ft above the surrounding ground to the east

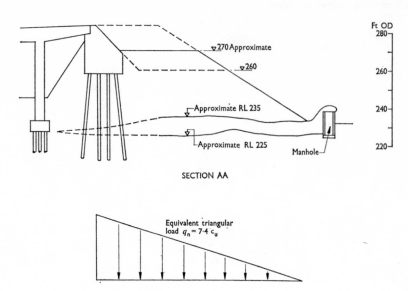

SECTION AA

Equivalent triangular
load $q_n = 7.4\, c_u$

Fig. 26

and about 35 ft above the motorway pavement. During construction of the sandy clay embankment a displacement of the fill occurred near the south bridge over a length of some 700 ft.

The intermittent lines show the boundaries of a former British Rail tip occupying the area of what was an earlier clay pit. The site investigation boreholes for the bridges passed through the tip straight into London Clay. A cross-section of the embankment and along section AA is shown in Fig. 26. The bridges on piled foundations and the lower parts of the fill had been completed for some time and the displacement occurred as additional fill was being raised around the back of the bridge abutment. Soft clay in a layer which subsequently proved to be 5–14 ft thick was squeezing out from beneath the stiff fill ($c_u \simeq 21$ lb/in.^{2}) and spreading in a plastic sheet over one of the manholes. Note the position of the manholes in Fig. 25 and on its cross-sections in Fig. 26.

Filling was immediately stopped and the embankment cut back just sufficiently to stabilize the movement. An intensive site investigation of the displaced slope and in the vicinity of both north and south bridge abutments was undertaken. Several piezometers installed at the base of the fill indicated pore-water pressures equivalent to a water height very considerably greater than the fill level.

The soft clay did not appear in the original survey boreholes and superficially it was similar to the London Clay. Consequently the embankment fill was placed directly on top of it. In fact the new survey suggested it was redeposited alluvial clay of no marked structure overlying the brown London Clay. Its strength however was considerably lower than that of the brown London Clay which was typically about 16 lb/in.^{2}

Analysis of the displacement suggested that the most likely explanation was a 'bearing' failure. The fill did not exhibit the characteristic vertical cracking and rotation of blocks associated with classical slides. Assuming an equivalent triangular loading as indicated in Fig. 26 the average undrained strength at the time of the slip was estimated at 3·4 lb/in.^{2}

To stabilize the embankment, consideration was given to natural drainage with regrading of the slope; drainage assisted by sand drains, and stone columns constructed by replacement techniques.

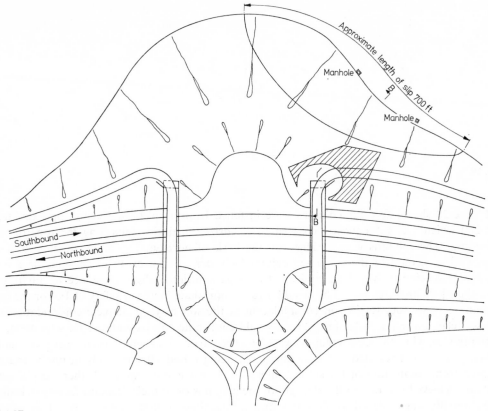

Fig. 27

Consolidation properties were measured on 3 in. diameter oedometer specimens cut both vertically and horizontally from samples. The soft clay was insensitive and very impervious with no noticeable difference in c_v values of about 4 ft²/year in the different planes, a value similar to that of London Clay. Its effective stress parameters were $C=1\cdot5$ lb/in.² and $\phi=17\cdot5°$.

It was estimated that the strength of the soft clay should be 8 lb/in.² for a factor of safety of 1·4 for the full 50 ft height of the embankment. To achieve the necessary strength by natural drainage under the existing 25 ft height of fill would require a 60% pore-pressure dissipation in the soft clay which would require up to 11 years according to drainage condition. Stability with nothing in hand would need about 1½ years. Regrading was impracticable because of the need to acquire additional land. Drainage through 1 ft diameter sand drains at as little as 5 ft centres would have required 1–2 years to achieve the same factor of safety.

Since the programme was critical, resort was made to stone columns which were installed within two months in the area shown in Fig. 27 where the height of all was potentially over-stressing the soft clay. The investigation had shown that the soft clay around the north bridge was somewhat stronger and no treatment was required there. To contain total cost and delay, the interchange was revised from an '0' plan shape to a 'U' shape. The height of embankment between the eastern ends of the bridges could then be reduced to a safe level as it exists today.

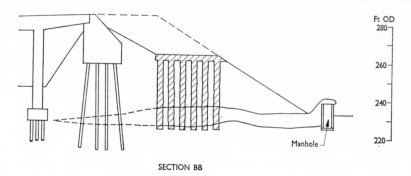

SECTION BB

Fig. 28

Figure 28 shows how the columns were installed from a level reduced to approximately RL 260 which was subsequently covered by a horizontal drain connecting the tops of the columns. The columns were designed on the basis of horizontal shear in the zone of soft clay, taking account of the contained column of water. 30 in. diameter columns on a 6 ft triangular spacing provided an equivalent average additional shear resistance of 3·6 lb/in.[2] over the whole area treated. This would give an immediate factor of safety of 1·3 for the full embankment without relying on consolidation of the soft clay. Complete consolidation of the soft clay under the embankment load would contribute approximately 5 in. settlement.

In practice, while the stone columns were being constructed, programme changes were made which in the event resulted in the embankment being raised after only about one year had elapsed from completion of the stone columns. Thus it is not known whether the ultimate stability achieved was due to the strength of the columns or to their drainage facility. Either way, stability was achieved within one year after installation of stone columns. This would not have been possible under natural drainage conditions unless the permeability of the clay had been underestimated by a factor of 10. Considering the lack of obvious structure in the clay this might seem unlikely but I leave the reader to decide which is the most likely explanation of the successful construction of the embankment.

I. Staal (Geotechnical Consultants Inc., Ventura, California) and K. Engelhardt (Vibroflotation Foundation Company, Pittsburgh, Pennsylvania)

Dr Engelhardt said that he would like to associate his remarks with those of Dr Staal, and submitted a joint contribution which is as follows.

Stone column foundations were selected for a 16 mgd sewage treatment plant located in a seismically active coastal area in California. 6524 stone columns were installed between December 1974 and May 1975. We wish to present data supplemental to prior field test results (see pp. 61–69), in particular the load–settlement relationship observed during the performance of 28 load tests on variable stone column patterns throughout the site. The contribution also comments on the relationship between load test results and anticipated total settlements for the structures.

Geotechnical conditions. The foundation soils consist of estuarine deposits; primarily inter-mixed sands, silts, and clays of low density. Older marine deposits consisting of dense silty sands underlie the estuarine soils. Depth of estuarine deposits range from 8 to 14 m. Ground water typically occurs within 1·5 m of ground surface.

Proposed plant. The plant is a conventional secondary treatment sewage disposal plant. The structures include sedimentation tanks, aeration tanks, digesters, thickeners, a chlorine contact tank, and sludge handling, maintenance and administration buildings. Concrete slabs and footings on a gravel distribution blanket will be used for foundations. Foundation loads vary considerably from structure to structure and also within the same structure. The general load range is from 0·5 to 2·0 kg/cm². The most significant structure is that of the biological units building comprising the sedimentation and aeration tanks. This structure has a total length of 180 m and a width of 75 m. The average foundation loading within the biological units is approximately 1 kg/cm.²

Stone column spacing and design loads. Square and rectangular patterns were selected for the layout of the stone columns. The densest pattern consisted of stone columns installed in a 1·22 m × 1·52 m rectangular pattern giving a foundation area of 1·85 m² per stone column. The most open pattern consisted of stone columns installed in a 2·13 m × 2·13 m square pattern giving a foundation area of 4·55 m² per stone column. All stone columns penetrated the soft estuarine soils and were founded in the underlying older marine sediments. Lengths of stone columns varied from 9 to 15 m. A series of stone columns was excavated and showed in-place diameters ranging from 0·80 m to 1·27 m, with an average diameter of 1·07 m.

Load tests. Twenty-eight vertical load tests were performed to evaluate the load–settlement behaviour. The test procedure was in accordance with the requirements of test method ASTM D 1194-72 (bearing capacity of soil for static load on spread footings), except that the standard 0·762 m diameter steel plate was replaced with circular concrete slabs of various diameters which were located concentrically over one stone column for each test. Each tested stone column was an interior column in a group of about 16 columns. The exact size of the circular concrete slab was determined by the stone column pattern in the area being tested. As an example, if the stone columns were constructed in a square grid pattern, 2 m × 2 m on centre, the circular concrete slab for the load test would have a diameter which would give the concrete slab a total area of 4 m².

The load was applied in 44·65 kN increments and no load increment was made before the rate of settlement was less than 0·25 mm/h. Maximum load was, in most cases, 357·2 kN. This load was sustained for a minimum of six hours after the 0·25 mm/h rate was reached prior to rebound. The load test was then rebounded. Maximum allowable settlement for each load test was 6·3 mm for the design load.

Prior to start of the load test programme it was decided to apply two different load test series, namely series A and series B. The purpose of series A tests was to establish that the soil conditions in a certain location were such that a stone column properly installed was able to carry the design load with less than 6·3 mm settlement. Rigid quality control, including measurements of rock take, vibration time and vibration energy was recorded for each individual stone column. If a series A load test indicated excessive settlement, an additional test in an adjacent area over the same pattern was performed. These tests were given a series C number. If the series C tests indicated excessive settlement and the quality control indicated proper workmanship, the stone column pattern was revised to a denser pattern to reduce the settlements. Load tests performed on such revised stone column patterns were designated series E.

A separate series of load tests designated series B tests was used to verify the contractor's performance in areas where the series A tests indicated anticipated geotechnical conditions. The series B load tests were often assigned to stone columns which had indicated a smaller rock take than the normal or a faster construction time than considered average for the general stone column installation. It was felt that by performing load tests on the most critical stone columns adequate load bearing capacity on the other stone columns would be ascertained.

It should be noted that most series A tests were performed within five days of the installation of the stone column, whilst most of the series B tests were performed three or four weeks after the installation of the stone columns.

Load–settlement curves. The load–settlement curves for the series A tests are shown in Fig. 29, and the load–settlement curves for the series B tests in Fig. 30. Fig. 31 compiles the curves for load tests indicating excessive settlements, subsequent retests on the same pattern, and load tests performed on other stone column patterns selected to improve the load–settlement relationship. The rebound curves have been deleted to avoid congestion of the figures. The design load is indicated by a dot on each individual load settlement curve.

The first load test indicating excessive settlements was load test 2A. This load test was performed on stone columns installed in a rectangular pattern 1·83 m × 1·98 m for a total area of 3·63 m² per stone column. A subsequent load test (2C) on a nearby stone column in the same pattern also indicated excessive settlement. The foundation soils in this area consisted of saturated soft silty clay. Based on the results of load tests 2A and 2C, combined with a review of the construction quality control information, it was decided that the spacing between stone columns in this area was too large for the soil conditions encountered. The stone column pattern was revised to 1·68 m × 1·83 m for a total area of 3·07 m² per stone column. A subsequent load test (2E) on the new pattern indicated acceptable load–settlement characteristics.

The next load test indicating excessive settlements was test 8A, which was performed on a stone column installed in a pattern spacing of 1·83 m × 1·83 m for a total of 3·34 m² per stone column. Excavation of the soils beneath the load plate after the test revealed that the load plate was underlain by 0·8 m of highly organic, uncompacted trash fill. The area and depth of the trash fill was determined by exploration. It was decided to remove and replace this unsuitable material prior to erection of the structure. To test the performance of the stone columns in the native material, test 8C was conducted beneath the trash fill. This load test indicated acceptable settlements without any revision of the stone column pattern.

Settlement against stone column spacing. The relationship between settlements and stone column pattern in square metres is plotted in Fig. 32. All settlements are plotted for a load of 268 kN (30 US tons). The graph indicates a fairly wide envelope for the relationship between settlement and pattern area. However, this relationship may be expected as the soil undoubtedly varies considerably from location to location.

Settlement against time. The time–settlement curve (Fig. 33) presents information regarding the consolidation characteristics of load test 11B under sustained load of 357·2 kN (40 US tons). The load was maintained for approximately five days. The test results were highly influenced by construction equipment working nearby and by temperature variations from day to night resulting in corresponding contraction or expansion of the steel reference beams, even though the beams were insulated by 25 mm thick Styrofoam plates.

Discussion. The first question that comes to mind after reviewing the load–settlement test results is the relationship between the settlements obtained from the load tests and the actual settlements resulting from loads imposed by large structures. Based on linear elastic theory, an analysis of stone column settlements has been made. This analysis considers a stone column as a compressible pile. Linear elastic material properties and moduli of elasticity were assumed as constant and independent of stress. The additional settlements resulting from loads on adjacent stone columns were evaluated by analysis of settlement of pile groups. The greatest uncertainty in this analysis is probably the assessment of the various moduli for the different earth materials. Also, the assumption of a constant modulus for the whole length of the stone columns can be questioned.

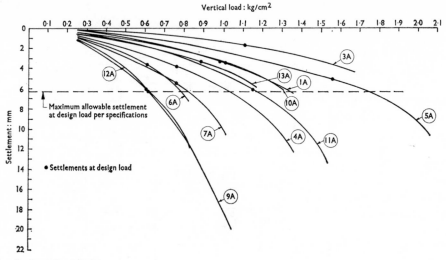

Fig. 29.   Series A load tests

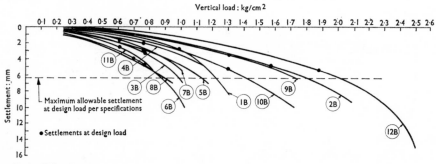

Fig. 30.   Series B load tests

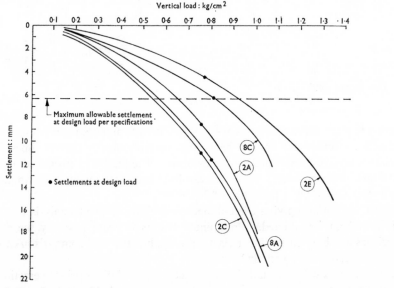

Fig. 31.   Failing tests and retests

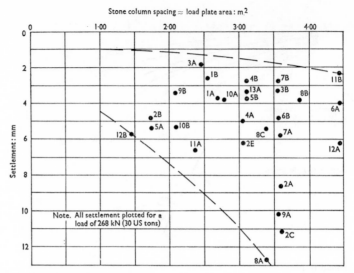

Fig. 32. Settlement against stone column spacing

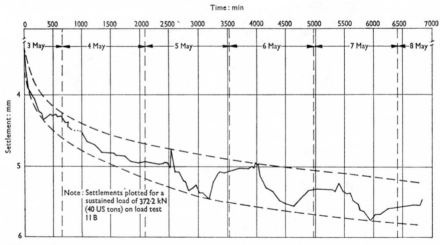

Fig. 33. Time against settlement

Assuming that the stone columns are supported on a firm base, the analysis suggests that the total immediate settlement should be approximatedly five to seven times the settlement measured in a load test on a single stone column.   The load tests produced an average settlement of 4 mm at design load.   Thus, the resulting total settlements after application of areal structural load should be in the range of 20–30 mm.

Additional research applying three-dimensional finite element analysis to a stone column group is now being conducted.   Preliminary results indicate basically the same settlement characteristics as obtained by linear elastic theory, but the final comparison can only be made after the study has been concluded.

Settlement gauges will be installed beneath the major buildings prior to erection.   Data from these settlement gauges will become available in 1976 and 1977.

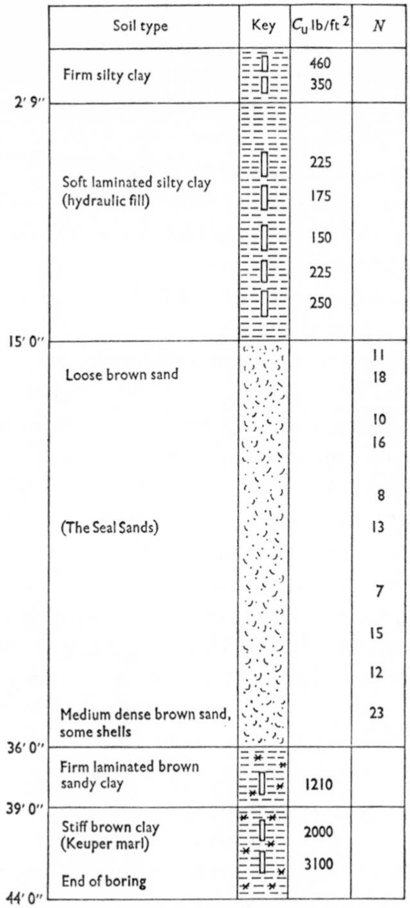

	Soil type	Key	C_u lb/ft^2	N
	Firm silty clay		460	
			350	
2' 9"				
	Soft laminated silty clay (hydraulic fill)		225	
			175	
			150	
			225	
			250	
15' 0"				
	Loose brown sand			11
				18
				10
				16
				8
	(The Seal Sands)			13
				7
				15
				12
	Medium dense brown sand, some shells			23
36' 0"				
	Firm laminated brown sandy clay		1210	
39' 0"				
	Stiff brown clay (Keuper marl)		2000	
			3100	
	End of boring			
44' 0"				

Fig. 34. Typical soil conditions

Conclusions. A series of 28 load tests performed during and after installation of stone columns for a major sewage treatment plant has verified anticipated load settlement characteristics. With proper workmanship and adequate spacing, modified as required by local soils conditions, the stone column improvement should provide an economical and adequate foundation for the structures.

Acknowledgements. We wish to thank Professor J. K. Mitchell of the University of California at Berkeley, for valuable information regarding analysis of settlements to stone column groups.

A. D. M. Penman (Building Research Station)

The limited experience we have had with stone columns placed under oil tanks at Teesmouth may be of interest. About 15 years ago ICI Ltd wished to develop a tank farm on reclaimed land consisting of 15 ft of pumped dredgings overlying the Seal Sands as shown in Fig. 34. The hydraulic fill was much too weak to support the weight of the proposed tanks (54 ft high) without treatment of some sort. The first four tanks were put on a reinforced concrete slab supported by Franki piles taken right through the Seal Sands into the underlying Keuper marl. These foundations were completely satisfactory, but rather expensive.

To effect an economy, the next seven tanks were placed on a Keller foundation consisting of stone columns at 9 ft centres. Each column was formed by grabbing a hole 6 ft × 3 ft in plan, taken down to the Seal Sand and backfilled with quarried blast furnace slag. This was compacted and forced sideways into the soft fill with a vibrating poker and further slag added to make up level. Finally, a pad of the quarried slag was formed over the columns and coated with a layer of asphalt to accept the base plates of the tank.

The remaining tanks used the BRS foundation which amounted to nothing except a ring of 14 in. diameter sand drains at 5 ft centres placed round the periphery of the tank, a pad of slag made thick enough to allow for the expected settlement and several instruments to monitor pore-pressures and settlements. By controlling the rate of initial loading to ensure that the applied stress did not exceed the increasing soil strength as consolidation occurred, the tanks were safely filled and the resulting settlements are given in Table 3. This table gives a comparison of cost and settlements for tanks of similar size on the three types of foundation. Unfortunately because the settlements of the Franki pile foundations were expected to be very small, no measurements of them were taken and also no gauges were placed under the tanks on the stone columns to measure central settlements. In general, edge settlements were measured at eight points round each tank and the greatest differential settlement given in Table 3 relates to the maximum difference between the settlements of these eight points and not the difference between edges and centres of the tanks. The table shows clearly that the more you pay the less the settlements, but the behaviour of the tanks with the simple foundation on this particular hydraulic fill has been entirely satisfactory.

Further details of this work have been given by Penman and Watson (1967).

S. Thorburn (Thorburn and Partners)

In the interesting paper by McKenna *et al.* (pp. 51–59) it is postulated that the stone columns were ineffective for two reasons. The grading of the 38 mm single size aggregate was too coarse to act as a filter, and the voids in the stone column became filled with clay slurry. The method of construction would probably have remoulded the soft clays and damaged the natural drainage paths. When forming stone columns in fine-grained soils it is advisable to use well graded hard natural aggregates ranging in particle size from 75 mm to 10 mm, otherwise contamination of the clean stone fill material can occur. Comments on precautions to be taken during the construction of stone columns have been given in the paper by Thorburn (p. 86).

The second reason postulated by McKenna *et al.* was clearly demonstrated by the large-scale field test at Grangemouth, the results of which are given in Fig. 2 of the paper by Thorburn (p. 87).

Table 3

Tank size		Foundation type	Approximate cost per ft² tank floor	Average edge settlement, ft	Greatest differential settlement, ft	Centre settlement, ft
Diameter, ft	Height, ft					
128	54	Franki piles	£2·20	—	—	—
128	56	Keller stone columns	£1·50	0·61	0·06	—
128	56			0·44	0·24	—
128	56			0·53	0·21	—
128	54	BRS foundations:	£0·40	0·89	0·13	1·36
186	54	pad and drains		0·71	0·10	1·02

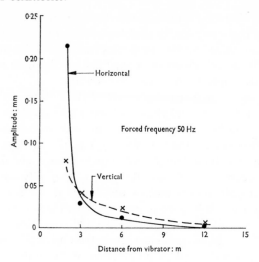

Fig. 35

It is of interest to examine the practical problem posed by the operation of depth vibrators in close proximity to existing buildings. A typical set of vibration measurements is presented in Fig. 35. The curves confirm the general experience of operators that an acceptable level of forced vibrations is induced at a distance of about three metres from traditional buildings. A difficult project was executed in the Bridgeton District of Glasgow some years ago where it was decided after discussion with the client to form stone columns within one metre of the external wall of the three-storey load-bearing masonry building. The existing foundations were disturbed by the construction of the stone columns to such an extent that old cracks in the masonry façade were widened. The disturbance caused by the ground stabilization work was, however, considered acceptable by the client and simple and relatively inexpensive renovation work sufficed to restore the appearance of the façade. With regard to vibro-compaction work in sand deposits the opinion of Dr Greenwood would be welcome in relation to the controlled cessation of jetting water from the nose cone of the depth vibrator at a distance of a few metres above the calculated depth of penetration of the nose cone. Is it possible that the jetting water will loosen the sand beneath the nose cone to such an extent that the subsequent ground treatment cannot restore the situation. This hypothesis may be important since the vertical components of the forced vibrations are relatively ineffective.

J. M. McKenna (*Consulting Geotechnical Engineer*)

I should like to reply to Dr Greenwood's comments on our paper (pp. 51–59). No one field experiment proves anything positively. The variability of natural ground deposits is such that we need a family of histories to be able to demonstrate something with reasonable assurance, so that our conclusions are not going to be proved wrong later. I got involved with the East Brent trial embankment after part of it had slipped and after the installation of the piles had been completed, and I came, therefore, fresh to the scene, having had no previous involvement in it. I found that the settlement records (p. 59) showed a distinct difference between the untreated end and the centre and the stone column end. The cross-section of the ground (Fig. 2, p. 52) shows that the simplified soil profile is a 12·5 m thick layer of soft clay and peat overlying grey silty sand. Dr Greenwood has pointed out that there are clay layers in this grey silty sand, which I accept, but, looking at the N values, one does get the impression of a reasonably competent bearing stratum.   Nevertheless, we have this paradox; the left hand end of the trial bank settled significantly less than either the centre or the right hand end, which was stone piled.   If the stone piles had been under the left hand end, I think the conclusion would have been drawn that they had significantly reduced the settlement! This emphasizes how misleading one set of results could be.

I think the key to what has happened is given by the results of the inductive settlement gauges (Figs 11 and 12, p. 59).   Fig. 11 shows values of settlement two days before the slip in the embankment, and it can be seen that the settlement of the upper 12·5 m was a uniform 440 mm.   The values 96 days after the slip, given in Fig. 12, show the same settlement in the upper 12·5 m, i.e. 650 mm at both ends of the embankment.   The difference in the surface settlement is therefore occurring at depth in the layer of grey silty sand, which I should have thought to be a material without very much settlement in it.   It is the settlement records of the 12·5 m thick clay layer which lead me to conclude that the stone columns were ineffective.

In the first place, if they had acted as drains, I should have expected the settlement in the stone column area to have been more rapid than in the untreated area, and the curve would therefore have flattened out earlier.   I see no evidence of this in the field records.

Second, Dr Greenwood is of the opinion that the stone columns did act as some form of support and that they failed by punching.   If this was the case, then before they reached the point of failure, the settlement in the stone column area would have been less than that in the unsupported area, and there is no evidence of this in the records.

Third, as Dr Greenwood has pointed out, the treated end was not involved in the slip.   It should be pointed out with equal emphasis that the untreated end was also not involved in the slip.   The fact is that the ends of trial embankments, where three-dimensional consolidation occurs, are more stable than the centres which can only consolidate two-dimensionally. Although the piezometer readings suggest that there was some load transfer in the stone column area, it is impossible to say how much the columns actually increased the stability. Stone columns have been used under other embankments in the eight years since the East Brent embankment was built.   It would be of great interest to the Profession if detailed records of their performance could also be published.

N. J. Withers (G. Maunsell and Partners)

By a stroke of good fortune I came across a reference (Moreau et al., 1835), during my studies at Cambridge, describing a series of experiments and projects, starting in about 1830, to determine how sand improved the bearing capacity of soft soils.   One of these projects used sand columns for the first time to support the heavy foundations of the ironworks at the artillery arsenal in Bayonne.   The military engineers had been compelled to try this method because wooden piles (the preferred solution) would have rotted in the estuarine deposits.

It was salutary to realize that the French had experimentally discovered that the piles transferred their load by arching to the side of the column and therefore that there was a maximum useful length.   This fact was not appreciated until my work confirmed it more than 130 years later.

They also suggested that the ultimate lateral stress the soil could withstand was equal to the bearing capacity of a surface footing.   Unfortunately they did not record any experiments to confirm this but it is interesting to compare this with my experimental results.   These showed that the ultimate lateral pressure can be approximated by $\sigma_{r1} = \sigma_{r0} + 4c$ in drained conditions, c is the undrained cohesion, σ_{r0} is the in situ lateral stress and σ_{r1} is the radial limiting stress.

In most normally consolidated soils σ_{r0} can be taken as $2c$ near the surface. Hence $\sigma_{r1} = 6c$ which agrees well with the French estimate of $6{\cdot}2c$.

However, the French were more concerned to reduce settlements. The columns used for the arsenal in Bayonne were two metres long 0·2 m in diameter and supported 10 kN each. They were constructed, on a progressively finer grid under the loaded area, by driving stakes into the ground, withdrawing them, then backfilling the holes with crushed limestone, until the stakes would not penetrate the ground more than 0·005 m under their standard hammer blow. Using this procedure they reduced settlements by a factor of four.

G. H. Thomson (*Cementation Ground Engineering*)

I should like to add a few comments to Mr Thorburn's contribution regarding the transmission of vibrations. I feel that it is unwise to accept any set of figures for the safe distance of operation of vibro-flotation from a structure as a general guide. Much depends on the actual ground conditions and the condition of the structure. We have records of two instances where old dwellings in excess of 10 m from the point of vibration were affected.

In the first instance at Lincoln some bricks fell from a chimney stack on a terraced house some 15 m away from the area being treated. The ground consisted of fill, principally sand and clay mixed with brick fragments, to a depth of 3 m overlying relatively dense sand. The houses were in very poor condition structurally, and were shortly to be demolished. In similar circumstances some bricks were dislodged from a chimney stack in a row of old terraced houses adjacent to a site on which vibro-flotation ground treatment was being undertaken at Wigan. These houses were also in a very poor state of repair with many loose bricks, and the bricks were dislodged when the point of treatment was just over 10 m away from the structure. The ground again consisted of fill, principally sand, ash, gravel and brick debris, generally about 3 m thick overlying sand and gravel.

The displacement amplitude measured on each site was 0·02 mm at a distance of 10 m from the centre of vibration with a frequency of 30 Hz, which is well below the displacement amplitude generally quoted as an upper limit for preventing cracking of brickwork, e.g. Morris (1950) quotes a maximum acceptable displacement amplitude of 0·4 mm.

Other cases have been noted where a structure some considerable distance from the site being treated has experienced noticeable vibration which has appeared to be due to the transmission of the vibrations along an inclined stratum of dense granular material outcropping beneath the foundations of the affected structure.

Transmission of vibrations can also be magnified by certain aspects of the particular structure. For instance, an old two-storey timber structure (considered to be of antiquarian interest) used as an antique shop had the first floor supported on steel props which were taken down through the ground floor and bedded into the underlying gravel. Vibrations from a vibroflot, well over 15 m away, were transmitted along the gravel stratum to the steel props and then to shelves attached to the props on which were antique china and glass objects. The shelves vibrated sufficiently to cause some of the objects to fall, although the general level of vibration within the structure was very low.

Whilst in general I agree with the curve shown by Mr Thorburn for homogeneous soils and most conventional sound structures, due consideration should be given to the state of the local structures, any peculiar features about them which may be significant in relation to possible vibration and to the stratification of the ground, particularly noting any dense granular stratum which could transmit vibrations over some distance.

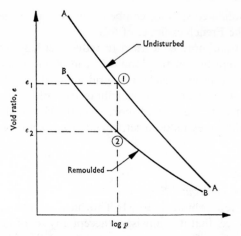

Fig. 36. Change in void ratio after remould if effective stress returns to original value

A. W. Bishop (*Imperial College, London*)

I wanted to raise one or two points about the philosophy underlying the use of stone columns. I am speaking as an outsider, not having used them, but I understand that we were discussing deep compaction. Having heard a number of people speak I now wonder if the users of these columns have not written off the soft soil as useless and are trying to replace as much of it as they can afford with something else, and are not really seriously considering compacting the clay in which the stone columns are made. On the other hand, if that were the case, it seems rather odd that they should put a vibrator down it and expend a great amount of energy hammering the stuff sideways when you could make the column do it more simply by some other means. I should like the exponents of stone columns to explain precisely what they think they are doing: are they just replacing the clay with something slightly better or are they trying to improve the ground around the stone columns by vibration or compaction in some sense and if so, what evidence have they that this has been achieved?

Now, if we consider diagrammatically a typical p–e curve (Fig. 36) for the undisturbed soil we might have a line such as AA, and for fully remoulded clay it might be indicated by BB. Assuming that we start off in the ground at point 1, in the vicinity of the stone columns, we will use a great deal of energy and we will probably fully remould the soil. So under the same normal effective stress we should expect the void ratio to drop from e_1 to e_2. This could mean a very substantial change in volume, so that if we were considering a stone column that had developed even a marginal zone of fully remoulded clay around it and less remoulded clay in between the columns, we should have to make up quite an extensive void ratio change by a volume change. This would take some time, unless we were working in a very silty material, and it will either be made up by settlement of the ground in between the columns or by a bulging of the stones and settlement of the column itself. It has got to be made up one way or the other as far as I can see: either way will cause a substantial decrease in void ratio adjacent to the columns, with a lesser decrease in between and if one put in a vane test or other type of apparatus of that sort, one should get a modest increase of strength here and a much more substantial increase immediately adjacent to the stone column (assuming an uncemented normally consolidated soil). There is already some evidence of this, but more would be of interest.

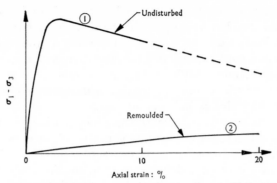

Fig. 37. Stress–strain curves for undrained tests on undisturbed and remoulded samples of normally or lightly overconsolidated clay

During discussion on a paper on Selset dam (Bishop and Vaughan, 1962) where we had used large diameter sand drains in a soft clay foundation, Professor Rowe suggested that one of the reasons why we obtained lower pore-pressures in the foundation was that the sand drains were supporting the weight of the dam. This seemed extremely unlikely because they would have punched into the case of the dam with the loads that would have to have been sustained. I did work out that with the density at which we were putting the crushed stone into the 18 in. diameter drains, the vertical strains would have required such a small mobilization of strength in the stone that it was quite compatible for the dam to settle as a unit, with very little differential settlement between the stone columns and the adjacent fill. Now again I would be interested at the sort of spacings that are used, to know how much evidence there is under actual foundations of load redistribution between the columns and the soft soil in between to show if extra load is in fact carried by the column.

It seems to me that there is a graduation between conventional sand drains at one end of the scale and what are now called stone columns at the other. If one is taking a slip surface through these stone columns, it appears that as soon as we start looking at the cross-sectional areas, one has to have a great deal of stone replacement to make very much impression on the strength because the angle of shearing resistance for the stone is only 40° or so, compared with 25° or 30° for a silty clay and 20° for even a fat clay; unless, of course, you have incredibly high pore-pressures in the clay in between. Now again, it is not apparent whether these are acting as sand drains: if they are you will not get these high pressures, so you do not need strength in the stone. If you need the strength of the stone, you need an awful lot of it in terms of percentage area; an amount which may be almost uneconomical to put in. Thus, are you replacing the clay or are you trying to strengthen it?

My own view is rather toward that of Dr Penman's or of the BRS, i.e. that it is best to do nothing if you can make the soil co-operate with you and use what strength it has. If one considers the strength characteristics of the soil and we take one of these soft varved clays as the undisturbed material, we can have a stress–strain curve as illustrated by curve 1 in Fig. 37. Incidentally, I was surprised to hear a sensitivity of 3 quoted; these soils seem to have a sensitivity closer to 10 when I have to deal with them. If you subject it to large strains the strength reduces to very low values. The remoulded clay has a stress–strain curve of the form illustrated by curve 2; it requires about 20% strain to reach full strength. If one reconsolidates it, even under the same pressure, in general one would not recover the same peak strength if one applied the same effective stress, but assuming that one did, one would do so at a strain that is very substantially greater than for the undisturbed clay. This strain would probably be 1 or 2% for the undisturbed clay, but would probably be 10% to reach the peak of the reconsolidated clay. If one is interested in deformations, this raises a problem. If the stone is considered as reinforcement it must be noted that the peak strength in dense stone columns will be reached at 3 or 4% axial strain. With 10–20% axial strain to the peak in clay one has an incompatibility in the properties of one's materials. This indicates a rather inefficient process of reinforcement from the philosophical point of view.

I seem to recollect that at Fornebu Airport where they put in a great many closely spaced driven sand drains in a rather sensitive clay, they got extremely large pore-pressures as soon as they drove them in, equal approximately to the overburden pressure. These dissipated rather fast initially, more slowly later on and I was going to enquire about the relative long-term settlements in sections where they drove sand drains at different spacings or omitted them. Some preliminary results were given by Eide (1963) at the European Conference at Wiesbaden.

I should like to go back to the general question: when is a stone column a stone column and not a sand drain—and if it is not a sand drain, is it really an effective way of reinforcing the ground? Why not either put in a pile or if one wants to use a stone column, why not just drive a mandril in the ordinary way—put some stones down it and pull the mandril out so saving all this disturbance of the adjacent clay which can only lead to further settlement and a greater lateral yield of the stone column than would have occurred if one had not disturbed the clay.

In conclusion, perhaps I could ask one question of Dr Penman which arose when he was discussing the Seal Sands, which I visited on one occasion. What were the properties of the clay overlying the sand: was it a silty clay with a fairly low clay fraction or low activity, or was it one of these fat, rather highly plastic clays that we get in the south of England in the Thames estuary, because I think there is a critical cut-off in the clays that one can usefully disturb and hope to strengthen by reconsolidation. On the other hand, for example, we have the Norwegian quick clays, which will liquefy when disturbed but will rapidly consolidate and then appear to be sustantially stronger than they were before. On the other hand, we have the Thames estuarine clays, which are less sensitive but will, nevertheless, undergo an 80–90% loss in strength on remoulding. However, their recovery in strength on consolidation is very much less useful from the engineering point of view, unless one is going to reconsolidate using a surcharge (which, I think, is something one would hope to avoid if one were using stone columns). A preliminary examination of the evidence suggests that there is a good deal still to be learnt about the effect of reconsolidation on the properties of remoulded, and partially remoulded (or disturbed) clays, in relation both to undrained strength and deformation modulus, and to drained strength and compressibility.

A. D. M. Penman (Building Research Station)

In reply to Professor Bishop, the hydraulic fill overlying the Seal Sands at Teesmouth was a silty clay with a fairly low clay fraction, i.e. a low activity which no doubt contributed to the satisfactory use of stone columns under some of the tanks. A comparison of the behaviour and costs of tank foundations using piles, stone columns and the simple BRS pad with encircling sand drains has been given in my earlier contribution and in greater detail by Penman and Watson (1967). The effect of stone columns was to halve approximately the edge settlement that occurred under tanks on the BRS foundation, although they made relatively little difference to the maximum differential settlement and their cost was about three times that of the BRS pad and sand drains.

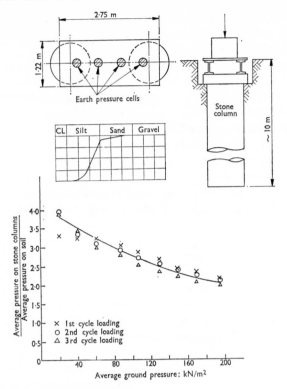

Fig. 38. St Helens: loading test for contact pressure

D. A. Greenwood (Cementation Limited)

   In reply to Professor Bishop's queries, I should make it clear that I do not aim to achieve compaction in clays between stone columns.   The mode of boring with water in soft potentially sensitive soils does not necessarily involve gross clay shearing but packing stone into the bore may produce a local skin of remoulded soil.   The vibration amplitude induced in the clay is very small and possibly many people have an exaggerated idea of the disturbance from this source.

   However, there is sometimes evidence of reduced clay strength immediately after column construction, increasing again to an enhanced value over a few days.   As might be expected, this is more often noticeable where vibro-displacement rather than replacement is being used in the stiffer, relatively insensitive, types of soil.   We have been collecting data about this for some years and hope to make it generally available when sufficient information has been accumulated.

   I do not endorse Professor Bishop's suggestion that columns might be formed as effectively and economically by ramming inside a casing.   It is probably not generally appreciated how efficient vibroflots are as boring machines and I do not like to contemplate potential problems of necking of a column of coarse frictional material involved in extraction of a casing.

   As to evidence of distribution of stresses between column and soil, we have made a number of field experiments, one of which is summarized in Fig. 38.   We have also studied, by means of earth pressure cells inside a column, the distribution of stress with depth; whilst predicted ratios of stress between cells are correct, we have difficulty in reconciling measured stresses with applied load and believe this to be a problem of experimental technique rather than a contradiction of current theories.

Turning to Mr McKenna's comments, I should like to clarify that the evidence of drainage I showed earlier (Fig. 23) for columns at East Brent, applies after construction and before loading. Because the column design was such that relative longitudinal movement took place between the column and soil, the warping of soil layers this induced may have sealed off the drainage as embankment load was applied.

I agree in principle with Mr Thorburn's remarks on the value of graded backfill for stone columns but better mechanical bond is of no avail if the stone does not flow freely into the bore. I prefer to try to ensure the latter. In any event, Mr Thorburn described the technique of lining the sides of the clay bore with stone which diminishes the need for a graded material since inter-penetration with clay has already been achieved to its physical limits and no further squeezing can occur.

J. B. Burland (Building Research Station)

I should like to reinforce some of the points made by Professor Bishop. In particular, I am concerned that so far the analyses that have been carried out on stone columns in clay soils have been based on considerations of undrained behaviour. As with the piling situation, we may get into serious trouble if we treat these materials as undrained when, in fact, they may be drained and we should be considering the behaviour in terms of effective stress.

With a stone column in a clay material, what probably happens in practice is that when the load comes on, from an oil tank or an embankment, the settlement in the central region will be fairly uniform, but the columns bulge. This will cause lateral compression of the clay which, in the long run, must be drained, contrary to the assumptions made by Hughes et al. (pp 31–44) which seem to be unrealistic during the working life of many structures founded on stone column foundations. The effectiveness of the stone column is critically dependent on the local horizontal compressibility of the soft soil. Yet the method of installation of these columns pays no regard to this and far from preserving the initial stiffness of the ground appears to result in significant remoulding. Compressibility, almost more than anything else, suffers with remoulding—a point emphasized by Professor Bishop.

I think it will be found that in some situations in particular types of ground with low sensitivity and where the disturbance is small, stone columns may well be sufficiently effective to be convincing, although I do not think that they can ever reduce the settlements by very large amounts in soft clays. In many other situations, however, particularly in the more sensitive materials, and where there is intense remoulding, I suspect stone columns will be totally ineffective. Therefore, until the behaviour of stone columns in soft clay is studied under fully drained conditions we are likely to accumulate a large amount of conflicting experience regarding their effectiveness.

N. J. Vadgama (Terresearch Limited)

It has been suggested that it is economic to use stone columns to transmit vertical loads through soft cohesive soils to stronger soils below. However, in our experience this may not be generally so because the bearing capacity of the stone columns is restricted by the limited lateral restraint provided by the soft cohesive soil. This is particularly the case when a group of stone columns is required to support a large base slab. For example, on a site where about 6·0 m of soft silty clay overlies a fine silty sand, the relative merits of using a raft foundation or stone columns may be looked upon as follows.

Adopt stone columns say 730 mm in diameter and use the relation (Hughes *et al.*, p. 33)

$$\sigma_v = \left(\frac{1+\sin\phi'}{1-\sin\phi'}\right)(4c+\sigma_{r'_o}) \qquad \cdots \cdots \cdots \cdots \quad (1)$$

Taking $\phi'=35°$, $c=25$ kN/m² and $\sigma'_{r_o}=20$ kN/m² for piles at edge of the footings, the ultimate bearing pressure that can be applied to the stone columns is:

$$\sigma_v = 3\cdot7[4(25)+20] = 444 \text{ kN/m}^2$$

Instead, using a square or circular raft foundation, the ultimate bearing pressure is:

$$\sigma_{vr} = (5\cdot5)(1\cdot2)(21) = 165 \text{ kN/m}^2$$

Hence

$$\sigma_v/\sigma_{vr} = 444/165 = 2\cdot7$$

Therefore for the stone columns to support the same ultimate load as the clay under a raft foundation, they would need to be spaced at about 1 m centres.

Thus, from load bearing considerations, using stone columns appears to be both unnecessary and impracticable. At the same time doubts have been expressed that there would be a general reduction in the rate and amount of settlement by using stone columns. In our limited experience we tend to agree with the adverse comments of McKenna *et al.* (pp 51–59) in this respect. On a large project near Hartlepool, we were asked to look into the possibility of using stone columns as an alternative to other forms of supporting a stockpile in an area underlain by soft to firm clay. Initially, in order to prove the validity of eqn (1) two samples taken from a borehole at the site were tested in a triaxial machine. The testing procedure consisted of drilling a 20 mm diameter hole in a 102×204 mm undisturbed sample and filling it with sand. The shaft of the plunger of the triaxial machine was used to apply a vertical load to fail the column. A rubber membrane was used to seal the sample and the inlet for the shaft in the platen, so that a confining pressure could be applied.

The results of the tests are summarized in Table 4, using a value of $\phi'=32\cdot5°$ for the sand columns in estimating σ_v from eqn (1).

The general ground conditions at the site were as shown in Fig. 39 and an estimate of the bearing capacity and spacing of 1 m diameter stone columns was made as below.

Assuming a radial stress approximately equal to the effective overburden pressure and considering the critical section at about 7·3 m depth $\sigma'_{ro}=2\cdot3(20)+5(10)=96$ kN/m². The cohesion of the silty clay was 40 kN/m². Adopting $\phi'=35°$ for the stone columns, the ultimate vertical stress is $\sigma_v=3\cdot7$ [4(40)+96]=947 kN/m². The estimated ultimate skin friction load on a stone column within the medium dense sand

$$(\phi' = 32°) = 0\cdot75[2\cdot3(20)+2\cdot5(10)]\times(\pi5\tan 32°) = 523 \text{ kN}$$

The estimated ultimate bearing capacity of a stone column $=523+[\pi/4(947)]=1267$ kN.

Table 4

Sample	c, kN/m²	σ'_{ro}, kN/m²	σ_v, kN/m²		$\dfrac{B}{A}$
			A Eqn (1)	B Actual	
1	35	138	923	965	1·05
2	16	103	555	520	0·94

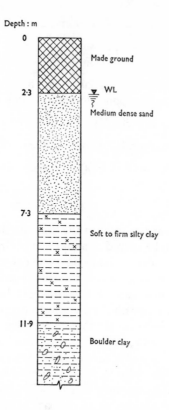

Depth : m

Made ground

WL

Medium dense sand

Soft to firm silty clay

Boulder clay

Fig. 39

Adopting $F=2$, the allowable load on a stone column $=633$ kN.   Average surcharge due to stockpile$=178$ kN/m². If all this load is transferred to the stone columns the equivalent area supported by one pile $=633/178=3.56$ m².

Therefore 1 m diameter stone columns would be required at 1·9 m centres each way.

Estimated costs of different types of foundation were compared and the use of stone columns was not an economic solution.   The method which showed most promise was the use of short-term preloading in the form of ground loading and dewatering.

P. Lubking, J. W. A. Jekel (*Delft Soil Mechanics Laboratory*) *and K. F. Brons* (*Nederhorst Grondtechniek BV*)

The case history as described by McKenna *et al.* deserves a thorough study.   The failure of the stone columns to act in the required fashion attributed to the filling of the voids in the gravel backfill with soft clay seems a logical conclusion.   Another defect of stone columns that is highlighted in the contribution of Hughes, *et al.* is the uncertainty in estimating the actual diameter of the column.

One deep compaction technique that overcomes the difficulty of controlling the diameter of the column and also allows sand to be used instead of stone, is the Compozersystem.   The use of sand will overcome the defect reported by McKenna *et al.*   As this process has not been used extensively in Europe it will be of interest to describe it in more detail together with the the results of one full scale test.

Ground compaction by the Compozersystem.   In 1957 a system of ground improvement was developed in Japan, which has been applied on a large scale for numerous projects.   The Compozersystem contains a number of aspects which compare favourably with the processes

Fig. 40.   Equipment used for Compozersystem

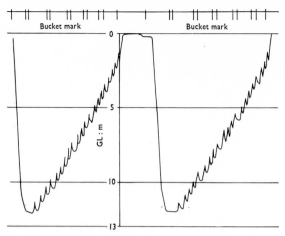

Fig. 41.   Movement recording of steel tube when making two Compozerpiles

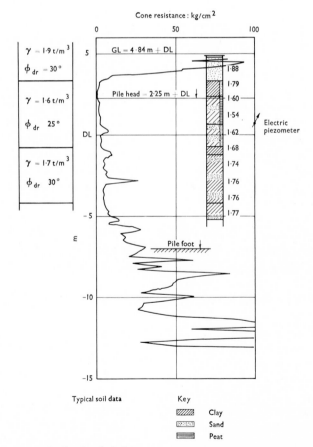

Fig. 42.   Dutch static cone penetration borehole log

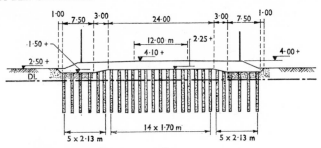

Fig. 43. Tank foundation with Compozerpiles: cross-section

available in Europe such as vibro-flotation and column compaction: it can be applied both in sandy and clayey soils; control of quantity of the added material is possible and it allows sand to be used under all conditions, resulting in cost saving in comparison with the use of gravel.

Compozersystem. The procedure in essence consists of sinking a thick-walled steel tube with a vibrator on the top into the subsoil to the required depth, if necessary with the aid of water and air jets. A sand plug at the lower end of tube prevents penetration of the soil into the tube. At the required depth a certain quantity of sand is placed inside the tube after which the tube is withdrawn over a predetermined depth, e.g. one metre. During the withdrawal sand flows out of the tube aided by air pressure that has been introduced on top of the sand. After withdrawal over this length the tube is sunk again partially, pushing the freshly deposited sand into the surrounding soil with the assistance of vibrations (Fig. 40). The ratio between the steps of extraction and redriving governs the cross-section of the sandpile. By repeating the steps gradually a pile is thus built up. The vibrations create a high density in the sand column. The groundwater table has no influence on the process. The sand to be used as fill must be clean and coarse graded. At the top of the tube a system of valves allows alternating introduction of sand or application of air pressure. A rigid quality control is imposed on the system by recording the parameters that govern the process, i.e. the quantity of sand added, the movements of the thick-walled tube and the power consumption of the vibrator (Fig. 41).

Design. The spacing and the cross-section of Compozerpiles is designed almost as for gravel columns constructed with vibroflots. The strength of the columns is governed by the angle of internal friction of the sand column and the lateral support of the original soil. The load is distributed over the sand piles and the original soil in the ratio of the relative compressibilities.

The first application in the Netherlands, and in fact in Europe, was for the foundation of four 36 m diameter 20 000 m^3 tanks. Allowable differential settlement between the centre and the perimeter was 30 cm; the maximum allowable distortion for the tank wall was 5 cm over 9 m (1:180). The maximum load during water-testing was ± 23 t/m^2 including the weight of the tank pad.

The subsoil consisted of sandy material from ground level at NAP+4·5 m to NAP+3·2 m; between NAP+3·2 m and NAP−2·0 m was a soft silty layer, placed hydraulically 10 years previously as a result of maintenance dredging works in the river Meuse. Between NAP−2 m and NAP−5 m there was clayey material with sandy lenses and deeper than NAP−5 m predominantly sandy material (Fig. 42). The groundwater table was found at NAP−1·5 m. The ground level had to be excavated to NAP+1·5−2·1 m as a base for the tank pad which had a thickness of 2–2·5 m.

Allowing for the original ground level at NAP+4·5 m, the settlement for the tanks without foundation improvement was calculated at 1·0–1·2 m for the tank centres and 0·6–0·8 m for the tank edges.

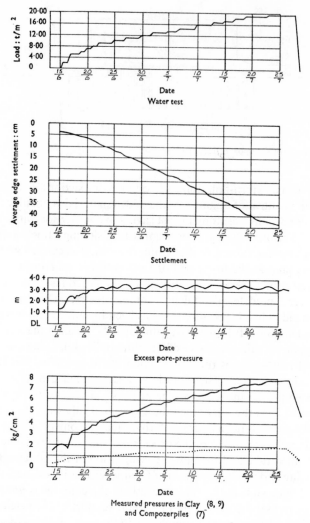

Fig. 44. Pressure distribution, excess pore-pressures and settlements during water testing

The design of the Compozerpiles aimed at reducing the settlements of the centre of 50–60 cm and of the edge to 30–40 cm. For the tank centre the area per pile varied between 2·5 m² and 3·5 m² and for tank edge 4 m² (Fig. 43). The quantity of sand introduced was designed to give a cross-section of the piles of 0·25 m² after compaction. The depth of piles was NAP– 7 m, hence just reaching the sandy strata. The allowable settlements amounted to approximately 50% of the settlements expected for the subsoil without treatment. Allowing for the compression of the untreated strata this means that the load on the compressible layers should not be more than approximately 10 t/m². It was computed that, allowing for the accelerating effect of the sand piles on the consolidation process, approximately 80% of the total settlements would take place during the water test (Fig. 44).

Measuring instruments were placed for two reasons: to check the pore-pressures so that the process of consolidation could be monitored; to check the assumptions made in the design stage in respect of load distribution between the sand columns and the natural clay.

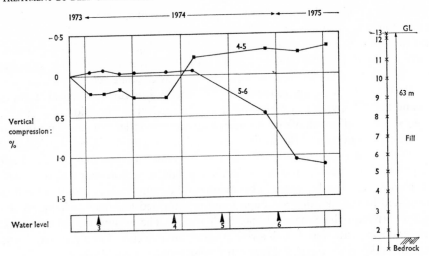

Fig. 45. Effect of a rising water table in a cohesionless fill: borehole settlement gauge B2

The measuring programme included: settlement beacons on the tank wall and in the tank centre; heave beacons along the toe of the tank pad; piezometers underneath the tank; pressure gauges on sand piles and between sand piles underneath the tank.

Acknowledgement. We acknowledge with thanks the opportunity provided by the client, Messrs Paktank, for the execution of the programme of measurements and the assistance during the water test of the instrumented tank.

OTHER MEANS OF COMPACTION

J. A. Charles (Building Research Station)

It is well known that certain types of unsaturated soil can undergo settlement on saturation. This suggests that a possible form of ground treatment for unsaturated fill materials in a loose state is inundation in which water is enabled to penetrate into the fill. Conversely if when building on such soils water gains access into the fill via drain trenches serious settlement can occur.

Ground treatment by inundation has been attempted at Corby on a site where the mainly cohesive fill left by opencast mining is about 24 m deep (see my earlier contribution). Trenches about one metre deep were dug at 10 m centres and kept full of water. Over a period of three months surface settlements of up to 14 cm have been measured. The vertical compression of the fill seems to have occurred mainly between about two and five metres depth. A programme of laboratory testing is being carried out in one metre diameter oedometers at BRS to determine the conditions under which the cohesive fill from Corby is liable to further settlement on saturation.

A different aspect of the effect of water on an unsaturated fill is being observed at a former opencast site where there is up to 65 m of predominantly cohesionless fill of shale, mudstone and sandstone fragments. Borehole settlement gauges of the magnet extensometer type used at Corby were installed with the co-operation of the NCB Opencast Executive and at this site settlements have been observed as the groundwater table rises in the fill. The borehole settlement gauges were installed while the pumping, which maintained a low water table, continued. Subsequently pumping ceased and the rising water level and the settlement of the ground are being monitored. Some of the movements that have been measured so far are illustrated in Fig. 45. At this settlement gauge during the first half of 1974 with the water table in the fill below the level of magnet 4, the fill between magnets 4 and 5 and between 5 and 6 showed little change in vertical strain. As the water table rose from the level of magnet 4 to the level of magnet 5 the fill between these two levels showed some vertical extension. With a further rise in the water table above the level of magnet 5 little further vertical strain occurred between 4 and 5, but between 5 and 6 the fill showed a vertical compression of over 1%.

Figure 46 shows similar behaviour monitored at another settlement gauge on the same site, although here the magnitude of the strains was smaller. Again significant vertical strains at a particular level in the fill occurred only as the water table rose through that level and again small extensions were measured at some levels and larger compressions at other levels. A possible explanation of the difference in behaviour at different levels in the fill as the water table rises is suggested by the fact that although loose unsaturated fills may be liable to settlement on saturation, a parent shale material may itself swell. The ground movements caused by a rising water table indicate that if such a rise in water table occurred at a restored opencast site subsequent to building on the site severe settlement problems could result.

D. N. Holt (*Freeman Fox and Partners*)

Dr Burland commented to the effect that in some of the applications of geotechnical processes to foundation engineering, practice has tended to develop ahead of theory. I should like to refer to a proposed application of a new method to a practical problem of road construction in South Wales. The method is similar to the dynamic consolidation process as it consists of dropping a heavy weight in a controlled and systematic manner. However, the application is in a totally different context, that of voided ground; and the object of the proposed application is also different, namely to provide stability by causing a controlled collapse of the uppermost and least stable parts of the void system. The problem has occurred in the design of a 1 km long section of the M4 where it skirts the southern rim of the South Wales coal basin. In the section in question, littoral facies of the Keuper series, consisting of a highly calcareous formation approximately 17 m thick, marly in the upper part and dolomitic in the lower part, overlie Coal Measures shales, sandstones, and thin coals. The dolomitic part is extensively voided, and carries groundwater in the form of a free flowing underground stream system, running across and below the line of the east–west oriented motorway. Right from the start of the project all the alarm bells were ringing for this area, in the form of comments on both the Ordnance Survey and the geological maps describing 'sinks', caverns and what were rather inappropriately referred to as landslides, and local lore about disappearing tractors and trailers. Nevertheless it became necessary for planning and environmental reasons to cross the area on an embankment up to 15 m in height, and so the designers were faced with finding practical solutions to the problems. The void system is not accessible for inspection so about 70 probe holes were sunk to ascertain its form and extent. This proved that the voids did not conform to any regular pattern, and that these occurred on several interconnected levels, of which only the lowest carried groundwater. Apart from open swallow holes, the shallowest voids found were 2·5 m below ground level and the deepest were at 17 m. Alternative remedial measures considered and rejected were as follows: construction without prior treatment and filling any voids created, but this was considered far too hazardous; injection of grout, to achieve the required selective closure of voids, but this was considered impractical, and the possible risk of obstruction of groundwater flow too great; provision of a soil cement, reinforced earth or polypropylene-strengthened base to the embankment, but the

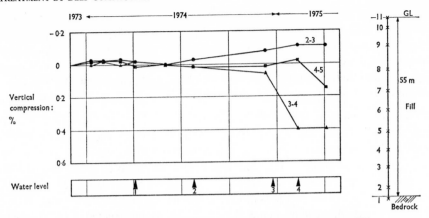

Fig. 46. Effect of a rising water table in a cohesionless fill: borehole settlement gauge D1

bridging effect that would be achieved was considered adequate only for small narrow voids; provision of a post-construction surcharge, but this does not avoid the greatest hazard, which is during construction, and is appropriate only to settlement rather than subsidence. Following these deliberations dynamic treatment seemed the only possible alternative, which these new approaches often tend to be. However it was felt that with control by careful monitoring of the ground by levelling and by probe drilling before and after treatment, and with careful exploration and survey of any voids actually opened up the method could be made to work. Generally depressions created would be treated by topping up with imported fill, but it was envisaged that any large breaches of the cavity system would be closed as would be done for a mine shaft, either with a purpose designed reinforcement concrete slab or mass concrete plug. It was also envisaged that where depressions occurred a number of passes would have to be made. Accordingly a specification was drafted covering all these proposals, and a bill of quantities prepared covering the work items envisaged, the end product being a complete sub-contract document covering a two-stage operation of trial and treatment. It has been said that an engineering design procedure was needed for the application of various new methods discussed here, and I would suggest that this document provides such an approach. The initiative for using these methods often comes from the contractor, but this is an example where the initiative has been with the engineer. In this context since the dynamic treatment is not necessarily only for compaction or consolidation, but can be for impaction, the term dynamic impaction is possibly more appropriate in such cases. In the normal course of events the work proposed would be now under way, but owing to changes in the road building programme it has been postponed, and I am unable to comment on the success of the method.

I should now like to refer to the appearance of gas bubbles at the surface following dynamic compaction, which has been quoted as evidence of soil compressibility being partly due to the presence of micro-bubbles of gases within the soil; I should like to ask how quickly these bubbles appear, since I feel that the appearance of such bubbles may be misleading. Pockets of air trapped between irregularities in the ground surface and the falling weight must inevitably be forced into the soil, but will almost immediately escape. The magnitude of the effect will vary with soil texture, presence or absence of vegetation, with topography, and with the shape of the falling weight. If the weight is provided with a raised edge or skirt as was described by Dr Engel, then the effect will be magnified many times. Has Mr Broise any comments on this? Finally, I should like to add a word of agreement on the conclusions expressed by Mr

McKenna regarding the use of single size stone in the columns below the Clevedon embankment. The need and the established practice of providing a purpose designed graded gravel screen to prevent silting in well construction seems an analogy worth quoting. It should also be mentioned that the common practice of using single size stone as a filter medium for drainage of highways and their embankments may often be as inappropriate as it seems to have been for the stone columns at Clevedon.

S. G. Shirke (Government of Maharashtra, India)

In India, when it comes to the construction of barrages on very wide rivers (sometimes more than a mile wide), the liquefaction of the sandy material in the foundation is an important factor in design considerations. The sand is generally found to be loose, fine and saturated. The firm rock may not be found until depths of 80–100 ft have been reached. Under these circumstances how far are the methods of deep compaction, vibro-flotation and so on effective and economical? Are there such schemes elsewhere already constructed or under construction?

D. W. Cox (Polytechnic of Central London)

The papers on deep compaction are very informative, yet there is no mention of the alternative of excavating fill and recompacting it in thin layers using a roller. This method is widely used for roads and earthworks and may be adapted to carry structures. The loading requirements for industrial and housing estates are generally similar to those for roads. Excavation and recompaction is often cheaper for fill depths to about 4–6 m. Excavation and recompaction to 2–6 m depths costs between £1·50/m² and £4·50/m².

Subsequent savings occur in floor slab construction and flexibility of layout. Whole areas are uniformly treated and factory or housing estates may be laid out to a scheme chosen, varied or extended at a later date. I have been involved in jointly designed schemes for nine structures where this method has been chosen and successfully completed at low cost. Regular settlement readings on fifty points on four structures over two years show maximum settlements of 20 mm and maximum differential settlements of 10 mm. These maxima are similar to the design values occurring when the standard penetration test is used on natural soils. Thorburn (1975) suggests that vibro-flotation methods are unsatisfactory where the soil contains unknown quantities of refuse, vegetable matter or voids due to arching of timber, concrete or steel drums etc. Dynamic consolidation may be similarly affected.

The alternative of excavation and recompaction enables small local deposits of particularly unsuitable waste to be removed from site or taken to the verges. Void formers such as beams, timber, drums, car bodies and even a pantechnicon (recorded beneath fill at Cardiff) can be removed. The remaining soil is mixed into a uniform deposit by the action of excavation and recompaction in thin layers. Granular wastes are most easily recompacted. Organic and unburnt refuse and soft clays are difficult to recompact by rolling unless they can be mixed in sufficient proportions with granular waste. Up to 30% average content of silt and clay may be acceptable. Excavation and recompaction by roller are not possible below the water table, but voids in fill placed at depth below the water table will normally have collapsed and compacted with wetting. The recompacted fill has an allowable bearing pressure of about 100 kN/m² but varying between 300 kN/m² for clean well graded granular soil and 50 kN/m² for sandy clays and silts. The values may be moderated by the depth of fill, and the condition of underlying natural strata.

The procedure with most sites has been a site investigation using test pits and exceptionally boreholes to examine and sample the fill. From this information a compaction specification has been prepared and on large contracts a compaction trial carried out. Site control is mainly by inspection, with rapid plate bearing tests on 600 mm diameter plates, and settlement tests on 1 m × 1 m pad footings loaded to 100 kN/m² and settlement recorded using a precise automatic level.   Scrapers and bulldozers are commonly used for excavating a long strip into which fill from an adjacent parallel strip is recompacted using a heavy vibrating roller.   Drainage layers may be incorporated if necessary.   Foundations may be pads or strips at shallow depth or surface rafts with carefully constructed joints, an edge beam, and tensile reinforcement beyond that used in road construction.   Continuous brickwork has been avoided where possible.   Compared with vibro-flotation the method has the advantage of being cheaper at shallow depth and more positive in enabling inspection and control of all the foundation soil. Since this exposes the soil an allowance which is not necessary with vibro-flotation must be made for wet weather.   Excavation and recompaction is relatively quiet, does not affect neighbours and does not normally require delivery of bulk materials to site.   The process is available from specialist contractors.

ECONOMICS
Y. M. Broise (Ménard Techniques Limited)

The question of economics is basically the most important one because it is the one that decides whether or not the method is used.   It is always raised when the method is being considered and it is one of the most difficult ones to answer because we are dealing with soils. Costs depend on the type of soil to be treated and the type of foundation to be placed on it. They are also dependent on the requirements for performance which are sometimes unrealistic, for instance when architects specify unnecessarily small settlements.   Every case has to be judged on its own merits, but looking back at our work in the UK, about 90% has cost between £2·50/m² and £3·50/m².   We have had various guideline prices which have been much higher: to satisfy certain conditions up to £6·00/m² or £7·00/m².   Some special things push up prices; a granular or good quality fill may be needed to form a working surface over soft clays, or the architect may want to make up for the enforced settlement that results from the treatment and the price of fill is usually high.   To make a blanket statement on costs would be wrong.

The process is, in general, applicable to areas in excess of 5000–10 000 m² if it is fairly free draining or fairly silty.   If it is clayey, it is practically impossible for us to look at anything under 15 000 m², unless there is a very low water table.   The process becomes very expensive for smaller areas on account of the pre-testing in the field and laboratory which has to be carried out before we can design the correct treatment.

To reply to Mr Holt, air trapped under the falling weight is a problem with a flat-bottomed tamper because the trapped air acts as a shock absorber.   This action is carried to the extreme with water: if the site is flooded, the layer of water acts as a shock absorber and dynamic consolidation cannot be carried out.   In order to minimize the effect, certain precautions are taken not to trap air.

G. H. Thomson (Cementation Ground Engineering Ltd)

I agree with Mr Broise, that it is very difficult to give general indications of cost because of the many factors that can affect the pricing, notably the type of structure, the required performance for that structure, ground conditions, size of job, and so on.   Nevertheless, guidelines are required for assessing general feasibility of ground treatment.   It should be appreciated that ground treatment is not technically directly comparable with a piled foundation but can be considered as an economical alternative.   In certain conditions, however ground treatment can offer a superior technical solution.

For conventional structures, comparison is usually made with piling and generally a saving

As compared to piles,

of the order of 25–40% can be made, taking into consideration all costs of piles, caps and beams as against conventional shallow foundations that can be used on treated ground.

For terrace houses as currently constructed in most urban redevelopment areas, the price per house unit can vary considerably, first on the size of the unit and second on the ground conditions and the depth of treatment necessary. However a range of £250 to £700 per house unit is common. Alternatively on the basis of cost per ton supported, which is a common yardstick used on piling schemes, ground treatment works out at around £0·50–£2·50 per ton with a range of treatment cost per linear metre around £6·00–£9·00.

The price per square metre is not normally used as a basis for comparison of price for vibro-flotation as, unlike dynamic consolidation which is essentially an area treatment, vibro-flotation can be used to provide support for specific foundation loadings, although it is also used for area treatment under embankments and for large loaded areas under oil tanks and warehouse floors. As a guide a range of £3·00–£6·00/m² can be used to evaluate possible cost of treating an area.

It should be appreciated that these are only very general prices which apply to the majority of projects, but there are, of course, the cases where particular specific conditions give rise to prices that fall outside this range.

To give an indication of some of the useful applications of treatment, which are not always readily apparent, ground contaminated with a high sulphate concentration, or which is highly acidic, possibly from an old gasworks or old chemical works, can be treated thus allowing the foundations to be kept. The cost of piling could be extremely high as permanent protective casing would be required in this instance. On large housing schemes all the foundations can be placed at uniform shallow depth after treatment, without the need for overdig for soft patches.

In conclusion I would emphasize its need for good quality site investigation, particularly of ground properties at shallow depths, before any real assessment of the viability of any form of ground treatment can be made.

REFERENCES

Bishop, A. W. (1966). Soils and soft rocks as engineering materials. *Inaug. Lect. Imp. Coll. Sci. Technol.* 6 289–313.
Bishop, A. W. & Vaughan, P. R. (1962). Discussion on Selset reservoir: design and performance of the embankment. *Proc. Instn Civ. Engrs* 23, 726–765.
Bratchell, G. E., Leggatt, A. J. & Simons, N. E. (1975). The performance of two large oil tanks founded on compacted gravel at Fawley, Southampton, Hampshire. *Proc. Conf. Settlement of Structures, Cambridge,* 3–9. London: Pentech Press.
Dash, B. P. & Hains, B. L. A. (1974). Moveout detection by autocorrelation matrix method. *Geophysics* V. 39, No. 6, 794–810.
Dash, B. P. & Obaidullah, K. A. (1970). Determination of signal and noise statistics using correlation theory. *Geophysics* V. 35, No. 1, 24–32.
Eide, O. (1963). Effects of vertical sand drains at Fornebu Airfield, Oslo. *Proc. European Conf. Soil Mech. Fdn Engng, Wiesbaden,* 95–97.
Greenwood, D. A. (1970). Mechanical improvement of soils below ground surface. *ICE Symp. Ground Engng,* 17, Fig. 11.
Hughes, J. M. O. & Withers, N. J. (1974). Reinforcing of soft cohesive soils with stone columns. *Ground Engng,* May, 42–49.
Marsland, A. & Quarterman, R. (1974). Further developments of multi-point magnetic extensometers for use in highly compressible ground. *Géotechnique* 24, No. 3, 429–433.
Meehan, R. L. (1967). The uselessness of elephants in compacting fill. *Canad. Geotech. Jnl* IV, No. 3, 358–364.
Moreau, Niel & Mary (1835). Fondations—emploi du sable. *Annales des Ponts et Chaussées.* Mémoires No. 224, 171–214.
Penman, A. D. M. & Watson, G. H. (1967). Foundations for storage tanks on reclaimed land at Teesmouth. *Proc. Instn Civ. Engrs* 37, Jan., 19–42.
Raffle, J. F. & Greenwood, D. A. (1961). The relation between the rheological characteristics of grouts and their capacity to permeate soil. *5th Int. Conf. Soil Mech. Fdn Engng, Paris* 2, 789.